A Guide
to
Topology

© 2009 by

The Mathematical Association of America (Incorporated)

Library of Congress Catalog Card Number 2009929077

ISBN 978-0-88385-346-7

Printed in the United States of America

Current Printing (last digit):
10 9 8 7 6 5 4 3 2 1

The Dolciani Mathematical Expositions
NUMBER FORTY

MAA Guides # 4

A Guide

to

Topology

Steven G. Krantz
Washington University, St. Louis

Published and Distributed by
The Mathematical Association of America

The DOLCIANI MATHEMATICAL EXPOSITIONS series of the Mathematical Association of America was established through a generous gift to the Association from Mary P. Dolciani, Professor of Mathematics at Hunter College of the City University of New York. In making the gift, Professor Dolciani, herself an exceptionally talented and successful expositor of mathematics, had the purpose of furthering the ideal of excellence in mathematical exposition.

The Association, for its part, was delighted to accept the gracious gesture initiating the revolving fund for this series from one who has served the Association with distinction, both as a member of the Committee on Publications and as a member of the Board of Governors. It was with genuine pleasure that the Board chose to name the series in her honor.

The books in the series are selected for their lucid expository style and stimulating mathematical content. Typically, they contain an ample supply of exercises, many with accompanying solutions. They are intended to be sufficiently elementary for the undergraduate and even the mathematically inclined high-school student to understand and enjoy, but also to be interesting and sometimes challenging to the more advanced mathematician.

MAA Service Center
P.O. Box 91112
Washington, DC 20090-1112
1-800-331-1MAA   FAX: 1-301-206-9789

Preface

Topology is truly a twentieth century chapter in the history of mathematics. Although it saw its roots in work of Euler, Möbius, and others, the subject did not see its full flower until the seminal work of Poincaré and other twentieth-century mathematicians such as Lefschetz, Mac Lane, and Steenrod. Topology takes the idea of non-Euclidean geometry to a new plateau, and gives us thereby new power and insight.

Today topology (alongside differential geometry) is a significant tool in theoretical physics, it is one of the key ideas in developing a theoretical structure for data mining, and it plays a role in microchip design. Most importantly, it must be said that topology has permeated every field of mathematics, and has thereby had a profound and lasting effect. Every mathematics student must learn topology. And physics, engineering and other students are now learning the subject as well.

Not only the content, but also the style and methodology, of topology have proved to be of seminal importance. Basic topology is certainly well-suited for the axiomatic method. Hence the flowering of the schools of Hilbert and Bourbaki came forth hand-in-hand with the development of topology. The R. L. Moore method of mathematical teaching was fashioned in the context of topology, and the interaction was just perfect. Modern gauge theory and string theory, and much of cosmology, are best formulated in the language of topology.

The purpose of this book is to give a brief course in the essential ideas of point-set topology. After reading this book, the student will be well-versed in all the basic ideas and techniques, and can move on to study fields in which topology is used with confidence and skill. We leave manifold theory and algebraic topology for another venue so that we can concentrate here on the most central and basic techniques in the subject. Students studying for qualifying exams will find this book to be a useful resource. Practicing mathematicians who need a place to look up a key idea can look here.

The book is written so as to be accessible and self-contained. There are many examples and pictures, together with timely discussion to put the key ideas in perspective. In just 100 pages the reader can come away knowing what topology is and what it is good for. We make an effort to hook topological ideas to more familiar concepts from calculus, analysis, and algebra so that the reader will always feel firmly grounded. Every concept in this book has a context. The book concludes with a treatment of function spaces, including the Ascoli-Arzela theorem and the Weierstrass approximation theorem. This seems to provide a fitting climax, and a nice set of applications, for what has gone before.

Since this is not a formal textbook, it does not have exercise sets. Most results in this book are proved in the traditional mathematical manner. We also include a Table of Notation and a Glossary in order to facilitate the reader's rapid acclimatization to the subject.

Because this book is in the nature of a handbook rather than a formal text, we have indulged in certain informalities. We sometimes will use a term or a concept before we have given its formal definition. We do so to avoid the sometimes cumbersome baggage of formal mathematics, and to keep the exposition light and accessible. In all such cases we make it clear from context what the concept means, and also make it clear where the more careful treatment of the term will appear. The reader should not experience any discomfort from this feature, and we hope will find that it makes the book easier to read.

We assume that the reader of this book has a solid background in undergraduate mathematics. This would of course include calculus and linear algebra, and a smattering of real analysis—especially the topology of the real line—would be helpful. To the extent possible, we have endeavored to fill in gaps so that minimal prerequisites are necessary. The typical (though not exclusive) reader of this book will be a graduate student studying for qualifying exams.

This is meant to be a book that can be read in a few sittings, just to get the sense of what this subject is about and how it fits together. It is different from a typical mathematics text or monograph. After reading this book (or even *while* reading this book), the reader may want to pick up a more traditional and comprehensive tome and work his/her way through it. Certainly, if one really wants to learn the subject, it is necessary to do plenty of exercises from those ancillary texts. The present book will serve as a good start on that journey.

This volume is part of a series by the Mathematical Association of America that is intended to augment graduate education in this country. As always, it has been a pleasure to work with the MAA and with Don Albers to develop the book. Underwood Dudley served as editor of this book series, and contributed many useful ideas and edits to help sharpen my exposition. The editorial board for this series also read the book extremely carefully and offered much wisdom and advice. Many of the figures throughout the book were produced by Ellen Klein and Julia Neidert, high school students in the University of Minnesota Talented Youth Mathematics Program under the direction of John Rogness. To all I am humbly grateful.

We hope that the present volume is a positive contribution to this new MAA book series.

St. Louis, Missouri   Steven G. Krantz

To
Bob Bonic

Contents

CHAPTER **1**

FUNDAMENTALS

1.1 WHAT IS TOPOLOGY?

In mathematics and the physical sciences it is important to be able to compare the *shapes* or *forms* of objects. Just what do we mean by "shape"? What does it mean to say that an object has a "hole" in it? Is the hole in the center of a basketball the same as the hole in the center of a donut? Is it correct to say that a ruler and a sheet of paper have the same shape—both are, after all, rectangles? What is a rigorous and mathematical means of establishing that two objects are equivalent from the point of view of shape or form?

1.2 FIRST DEFINITIONS

A *topological space* is a set X together with a collection of subsets $\mathcal{U} = \{U_\alpha\}_{\alpha \in A}$ that we call the *open sets*. We assume that

(a) The entire space X is open.

(b) The empty set $\emptyset$ is open.

(c) If U_β are the open sets then $\bigcup_\beta U_\beta$ is another open set ($\mathcal{U}$ is closed under the union operation).

(d) If U and V are open sets then $U \cap V$ is an open set ($\mathcal{U}$ is closed under pairwise, indeed finite, intersection).

The entire subject of topology is based on the idea that if you know the open subsets of a space then you know about its form. We sometimes write our topological space as $(X, \mathcal{U})$.

EXAMPLE 1.2.1. Let $X = \mathbb{R}$, the familiar real number system. Let us declare every open interval (a, b) to be an open set (here we allow a to be $-\infty$ or b to be $+\infty$). We also declare any union of such intervals intervals to be an open set. It is easy to see that the collection $\mathcal{U}$ of all the open sets that we have described forms a topology (of course the empty set is the union of *no* open intervals). The entire real line is an open set according to our definition. Typical open sets are

$$U_1 = (-1, 3) \cup (4, 9) \cup (16, 31)$$

and

$$U_2 = \cdots \cup (-3, -2) \cup (-1, 0) \cup (1, 2) \cup (3, 4) \cup \cdots$$

and

$$(-\infty, -5) \cup (7, +\infty).$$

We call this topology the *standard topology* or the *usual topology* or sometimes the *Euclidean topology* on the real line.

EXAMPLE 1.2.2. Let $X = \mathbb{R}^N$, the standard Euclidean space of N dimensions. We use coordinates to denote a point $x \in X$ by $x = (x_1, x_2, \ldots, x_N)$. Let us say that U is an open set if, whenever $x \in U$, then there is an $\varepsilon > 0$ such that the ball $B(x, \varepsilon) \equiv \{t \in \mathbb{R}^N : |t-x| < \varepsilon\}$ lies in U. We commonly refer to $B(x, \varepsilon)$ as the *open ball* with center x and radius ε.

The entire space $U = \mathbb{R}^N$ is an open set according to this definition, and so is the empty set. But there are lots of other open sets as well. The set $T = \{t = (t_1, t_2, \ldots, t_N) \in \mathbb{R}^N : |t_j| < 1 \text{ for } j = 1, 2, \ldots, N\}$ is easily checked to be an open set. For if $x = (x_1, \ldots x_N) \in T$, then let $\varepsilon = \min\{|x_j \pm 1| : j = 1, \ldots, N\}$. Then $B(x, \varepsilon) \subseteq T$ as required. See Figure 1.1.

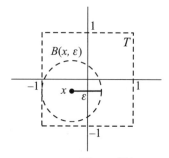

FIGURE 1.1. The set T is open.

The open sets are closed under the union operation and under finite intersection. So $\mathbb{R}^N$, equipped with the notion of openness specified here, is a topological space.

We call this topology the *standard topology* or the *usual topology* or sometimes the *Euclidean topology* on $\mathbb{R}^N$.

It is worth noting that the topology described here is "generated" by all the balls $B(x, \varepsilon)$. Each such ball is open, just as the example set T above was open. Secondly, a union of such balls is clearly open. Third, if U is any open set according to our definition in the first paragraph, then each point in U has a ball about it (that lies in U). So U is a union of open balls. Thus the collection of unions of open balls is just the same as the collection of open sets.

It is also worth observing explicitly that the topology described here in case the dimension N is 1 is just the same as the topology considered in the last example.

Remark 1.2.1. For the record, we note that the *closed ball* with center x and radius ε in Euclidean space is given by

$$\overline{B}(x, \varepsilon) = \{t \in \mathbb{R}^N : |t - x| \le \varepsilon\}.$$

EXAMPLE 1.2.3. Let X be the unit interval $[0, 1]$ and let the only open sets be the empty set $\emptyset$ and the entire interval $[0, 1]$. One may check directly that all the axioms for a topological space are satisfied.

The last example—which we sometimes call the *trivial topology*—will work on any non-empty set X.

EXAMPLE 1.2.4. Let X be the set of integers $\mathbb{Z}$. Call a set open if it is either empty or all of $\mathbb{Z}$ or is the complement of a finite set. Then it is straightforward to confirm that this is a topological space.

We may note that the last example works for any infinite set X (not just the integers $\mathbb{Z}$).

EXAMPLE 1.2.5. Consider the topology on the real line generated by intervals of the form $[a, b)$ or $[a, +\infty)$ (here we mean "generated" in the sense of taking finite intersection and arbitrary union). This is called the *Sorgenfrey line*, named after Robert Sorgenfrey (1915–1996). The Sorgenfrey line is one of the most important examples in topology.

First note that, if (c, d) is any open interval, then

$$(c, d) = \bigcup_{\varepsilon > 0} [c + \varepsilon, d).$$

Thus (c, d) is the union of Sorgenfrey open sets. So any standard open interval is open in the Sorgenfrey topology. We see, then, that the Sorgenfrey topology contains all the usual open sets and some new ones as well.

We shall encounter the Sorgenfrey line at several junctures in the sequel; it is an important example in several contexts of point-set topology.

If $(X, \mathcal{U})$ is a topological space then $(Y, \mathcal{V})$ is a *topological subspace* (or more simply a *subspace*) if $Y \subseteq X$ and each $V \in \mathcal{V}$ is of the form $V = Y \cap U$ for some $U \in \mathcal{U}$.

In a topological space $(X, \mathcal{U})$, a set $E \subseteq X$ is called *closed* if its complement $X \setminus E$ is open. In Example 1.2.1, any interval $[a, b]$ is closed (though these are certainly not all the closed sets!—see Section 1.11 on the Cantor set). In Example 1.2.3, the only closed sets are the entire interval $[0, 1]$ and the empty set. In Example 1.2.4, the closed sets are the finite sets (and the empty set and the entire space X).

There are sets that are neither open nor closed, such as the interval $[0, 1)$ (Example 1.2.1). It is also possible for a set to be both open and closed. In any topological space $(X, \mathcal{U})$, the entire space X and the empty set $\emptyset$ are both open and closed.

Proposition 1.2.2. *The union of two closed sets is closed.*

Proof: Let the two closed sets be E and F. Then $X \setminus E$ and $X \setminus F$ are open. So
$$S \equiv (X \setminus E) \cap (X \setminus F)$$
is open. But then
$$^c S \equiv X \setminus S = E \cup F$$
is closed.   □

Proposition 1.2.3. *Let $\{E_\beta\}_{\beta \in B}$ be closed sets. Then $\cap_\beta E_\beta$ is also closed.*

Proof: Exercise for the reader.   □

We want to develop a language for describing and analyzing the parts of a set. Consider the closed disc $\{(x, y) \in \mathbb{R}^2 : x^2 + y^2 \leq 1\}$ depicted in Figure 1.2a. We can see intuitively that this set has a boundary (the circle—see Figure 1.2b). And it has an interior (the open disc $\{(x, y) \in \mathbb{R}^2 : x^2 + y^2 < 1\}$—see Figure 1.2c). We would like a precise description of the boundary and interior of *any* set. For example, consider the set S of integers in the topological space $\mathbb{R}$ with the usual topology, described in Example 1.2.1. What is its interior and what is its boundary? The answer to this last question is not obvious, and requires some study.

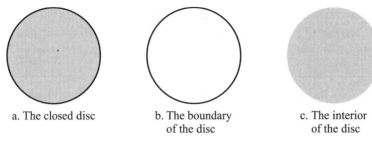

a. The closed disc

b. The boundary
of the disc

c. The interior
of the disc

FIGURE 1.2.

First we need a bit of terminology. If X is a topological space and $x \in X$ then a *neighborhood* U of x is an open set that has x as an element.

Definition 1.2.6. Let S be any set in the topological space X. A point $p \in X$ is said to be an *interior point* of S if

(a) $p \in S$,

(b) There is a neighborhood U of p such that $p \in U \subset S$.

Definition 1.2.7. Let S be any set in the topological space X. A point $q \in X$ is said to be a *boundary point* of S if any neighborhood U of q intersects both S and $^c S \equiv X \setminus S$.

These new concepts can be elucidated through some examples.

EXAMPLE 1.2.8. Let us return to the example of the integers $\mathbb{Z}$ as a subspace of the real line equipped with the Euclidean topology. If p is any point of $\mathbb{Z}$ and U is any neighborhood of p then it is clear that U will contain points that do not lie in $\mathbb{Z}$. See Figure 1.3. Therefore there is no point that is an interior point of $\mathbb{Z}$, so the interior of $\mathbb{Z}$ is empty.

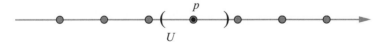

FIGURE 1.3. The point p is in the boundary of $\mathbb{Z}$.

If q is any point of $^c \mathbb{Z}$ then let $\varepsilon > 0$ be the distance of q to the nearest integer. See Figure 1.4. Then the interval $V = (q - \varepsilon, q + \varepsilon)$ is a neighborhood of q that intersects $^c \mathbb{Z}$ but does not intersect $\mathbb{Z}$ itself. So q cannot

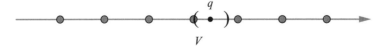

FIGURE 1.4. The point q is not in the boundary of $\mathbb{Z}$.

be a boundary point of $\mathbb{Z}$. If instead q is an element of $\mathbb{Z}$ and U is any neighborhood of q then $q \in U \cap \mathbb{Z}$ and, plainly, U will also contain nearby points that are not in $\mathbb{Z}$ (see Figure 1.3). Thus q is a boundary point of $\mathbb{Z}$. We see then that the boundary of $\mathbb{Z}$ is just $\mathbb{Z}$ itself.

If S is a set in a topological space X then we denote the interior of S by $\overset{\circ}{S}$ and the boundary of S by ∂S.

EXAMPLE 1.2.9. Consider X the real line with the usual topology (as in Example 1.2.1). Let S be the set $[0, 1)$. Then the boundary of S is the pair $\partial S = \{0, 1\}$, and the interior is the interval $\overset{\circ}{S} = (0, 1)$.

EXAMPLE 1.2.10. For the topological space in Example 1.2.3, let S be any proper subset of the interval $X = [0, 1]$. If p is any point of S then the only possible neighborhood of p is the entire interval $[0, 1]$ because that is the only open set available. Since that interval cannot lie in S, we conclude that S has no interior points.

Now if q is *any point* in the entire space X then the only neighborhood of q is the entire interval $[0, 1]$ (since that is the only open set available). And that interval intersects both S and its complement. We conclude that ∂S is the entire space $[0, 1]$.

The interior and the boundary of a set have many interesting properties. We shall record a few of them here. We begin with a lemma that has independent interest.

Lemma 1.2.4. *Let S be a set in a topological space X. If each point $s \in S$ has a neighborhood that lies in S, then S is open.*

Proof: Let $s \in S$ and let U_s be the neighborhood of s that lies in S. We have

$$S = \bigcup_{s \in S} \{s\} \subseteq \bigcup_{s \in S} U_s \subseteq S .$$

We conclude that

$$\bigcup_{s \in S} U_s = S .$$

But the set on the left of this last equality, being the union of open sets, is open. Hence S itself is open. $\square$

Proposition 1.2.5. *The interior of any set S is open.*

Proof: If $p \in \overset{\circ}{S}$ and U is a neighborhood of p that lies inside S then any point x of U is also in the interior of S. For U will be the required neighborhood of x that lies in S.

Now let $S \subset X$ be any set and let p be a point of its interior. We know that there is a neighborhood U_p of p such that U_p lies entirely in S. We then know, by the last paragraph, that $U_p \subseteq \overset{\circ}{S}$. Therefore

$$\overset{\circ}{S} = \bigcup_{p \in \overset{\circ}{S}} \{p\} \subseteq \bigcup_{p \in \overset{\circ}{S}} U_p \subseteq \overset{\circ}{S}.$$

We conclude that

$$\bigcup_{p \in \overset{\circ}{S}} U_p = \overset{\circ}{S}.$$

The set on the left-hand side of this last expression, being the union of open sets, is open. Therefore $\overset{\circ}{S}$ is open. $\square$

Proposition 1.2.6. *The boundary of any set S is closed.*

Proof: Let x be a point that is not in ∂S, the boundary of S. Then there is some neighborhood U of x that does not intersect both S and $^c S$. It follows that any point $t \in U$ also has such a neighborhood, namely U itself. So U lies in the complement of ∂S. It follows, by the lemma, that the complement of ∂S is open. So ∂S is closed. $\square$

Definition 1.2.11. The *closure* of a set S is defined to be the intersection of all closed sets that contain S. We denote the closure of S by $\overline{S}$.

Of course the closure $\overline{S}$ of S is closed.

Proposition 1.2.7. *The set $\overline{S}$ equals the union of S and ∂S.*

Proof: Suppose that x is a point that is not in $S \cup \partial S$. Since x is not in ∂S there is a neighborhood U of x that either does not intersect S or does not intersect $^c S$. We know that $x \notin S$, so it must be that $U \subseteq {}^c S$. So in fact we see that every point of U is not in $S \cup \partial S$. Thus the complement of $S \cup \partial S$ is open, and $S \cup \partial S$ is closed. We conclude that $S \cup \partial S \supseteq \overline{S}$.

If instead $x \notin \overline{S}$, then, since $^c \overline{S}$ is open, there is a neighborhood U of x that lies in $^c \overline{S}$ and hence in $^c S$. So certainly U is disjoint from $S \cup \partial S$ and thus, in particular, $x \notin (S \cup \partial S)$. We conclude that $S \cup \partial S \subseteq \overline{S}$.

The two inclusions taken together give our result. $\square$

The last proposition is consistent with our intuition. In Figure 1.2a, the closed disc, we see that the closure is simply the closed disc, which is the interior plus the boundary circle. In other words the closure of S is the interior set $\overset{\circ}{S}$ union its boundary ∂S.

A commonly used term in this subject is "accumulation point." If $(X, \mathcal{U})$ is a topological space and $A \subseteq X$, then we say that x is an *accumulation point* of A if every neighborhood of x contains points of A other than x itself. As an instance, if $X = \mathbb{R}$ with the Euclidean topology and $A = (0, 1)$, then the points 0 and 1 are accumulation points of A. In fact the set of *all* accumulation points for A is just the closed interval $[0, 1]$. We leave it to the reader to use the ideas presented here to show that a closed set in a topological space contains all its accumulation points. We can actually say more, and we state it as a formal proposition:

Proposition 1.2.8. *Let $(X, \mathcal{U})$ be a topological space and $S \subseteq X$. Then the closure of S equals the union of S and its accumulation points.*

Proof: Exercise for the reader. Imitate the method of Proposition 1.2.7.

$\square$

If X is any set then the topology just consisting of X itself and the empty set $\emptyset$ is the smallest topology on X. By contrast, the topology in which each singleton $\{x\}$ for $x \in X$ is open is the largest topology on X. We call the latter topology the *discrete topology*, and we say that the space is discrete.

1.3 MAPPINGS

Our principal tool for comparing and contrasting topological spaces will be mappings. The mappings that carry the most information for us are the continuous mappings. A *mapping $f : X \to Y$* is a function on a set X that takes values in a space Y rather than in the real numbers or the complex numbers. We shall formulate our notion of continuity in terms of the inverse image of a mapping. Let $f : A \to B$ be a mapping. Let $S \subseteq B$. Then $f^{-1}(S) \equiv \{x \in A : f(x) \in S\}$. We call this set the *inverse image of the set S under the mapping f*.

Definition 1.3.1. Let $(X, \mathcal{U})$ and $(Y, \mathcal{V})$ be topological spaces. A function (or mapping) $f : X \to Y$ is said to be *continuous* if, whenever $V \subseteq Y$ is open, then $f^{-1}(V) \subseteq X$ is open.

Remark 1.3.1. This definition requires some discussion. We are working in an abstract topological space. We do not necessarily have a notion of distance, so we cannot say "if the variable x is less than δ distant from c then $f(x)$ is less than ε distant from $f(c)$." We instead rely on our most fundamental structure—the open sets—to express the idea of continuity. We will have to do some work to see that the new notion of continuity, in the appropriate context, is equivalent to the old one.

Remark 1.3.2. Let us note here the rigorous definition of continuity that we learn in calculus and real analysis:

Let I be an open interval in the real line and $f : I \to \mathbb{R}$ a function. Fix a point $c \in I$. We say that f is *continuous* at c if, for any $\varepsilon > 0$, there is a $\delta > 0$ such that, whenever $|x - c| < \delta$, then $|f(x) - f(x)| < \varepsilon$.

This definition, thought about properly, makes good intuitive sense.

Proposition 1.3.3. *On the real line, the definition of continuity in Remark 1.3.2 is equivalent to Definition 1.3.1 (formulated in the language of inverse images of open sets).*

Proof: Suppose that I is an interval in $\mathbb{R}$ and $f : I \to \mathbb{R}$. Assume that f is continuous according to the classical definition in 1.3.2. Let V be an open subset of $\mathbb{R}$; we must show that $f^{-1}(V)$ is open. Consider the set $f^{-1}(V)$ and let $x \in f^{-1}(V)$. Then $f(x) \in V$ and, since V is open, there exists an $\varepsilon > 0$ such that the interval $(f(x) - \varepsilon, f(x) + \varepsilon) \subseteq V$. By the classical definition, there exists a $\delta > 0$ such that if $t \in (x - \delta, x + \delta)$ then $f(t) \in (f(x) - \varepsilon, f(x) + \varepsilon)$. This says that the interval $(x - \delta, x + \delta)$ lies in $f^{-1}(V)$. Hence $f^{-1}(V)$ is open. We have shown that the inverse image of an open set is open, and that is the new definition of continuity.

For the converse, assume that $f : I \to \mathbb{R}$ satisfies the new definition of continuity given in Definition 1.3.1. Fix a point $x \in I$. Let $\varepsilon > 0$. The interval $(f(x) - \varepsilon, f(x) + \varepsilon)$ is an open subset of the range $\mathbb{R}$. Thus, by hypothesis, the inverse image $f^{-1}((f(x) - \varepsilon, f(x) + \varepsilon))$ is open. It is an open neighborhood of the point x. So there exists a $\delta > 0$ such that $(x - \delta, x + \delta) \subseteq f^{-1}((f(x) - \varepsilon, f(x) + \varepsilon))$. This means that if $|t - x| < \delta$ then $|f(t) - f(x)| < \varepsilon$, which is the classical definition of continuity given in 1.3.2. $\square$

One thing that is nice about our new definition of continuity is that it is simple and natural to use in situations where the traditional definition would be awkward. We examine some examples.

EXAMPLE 1.3.2. Let $f : \mathbb{R} \to \mathbb{R}$ be given by $f(x) = x^2$. Discuss the continuity of f.

We know from experience that f is continuous—all polynomial functions are continuous. But it is instructive to examine the new definition of continuity in this context.

Let V be an open subset of the range space $\mathbb{R}$. We may take V to be an interval $I = (a, b)$ since any open set is a union of such intervals (exercise). Then

- If $0 \leq a < b$ then $f^{-1}(I) = (\sqrt{a}, \sqrt{b})$, and that is an open set.

- If $a < 0 \leq b$ then $f^{-1}(I) = (0, \sqrt{b})$, and that is an open set.

- If $a < b < 0$ then $f^{-1}(I) = \emptyset$, and that is an open set.

We have verified directly that f is continuous according to the new definition.

EXAMPLE 1.3.3. Let us equip the rational numbers $\mathbb{Q}$ with the topology that the number system inherits from the superset $\mathbb{R}$. This means that a set $V \subseteq \mathbb{Q}$ is open precisely when there is an open set $U \subseteq \mathbb{R}$ such that $U \cap \mathbb{R} = V$.

Consider the function $f : \mathbb{Q} \to \mathbb{Q}$ that is defined as follows. If p/q is a rational number expressed in lowest terms (i.e., p and q have no prime factors in common), with q positive, then set $f(p/q) = 1/q$. Determine whether f is continuous at any point.

In fact f is discontinuous everywhere. The values of f are $1/1$, $1/2$, $1/3$, $1/4$, etc. Let us take a neighborhood V of $1/2$ in the image—this is a typical open set. We take the neighborhood to be an interval that in $\mathbb{Q}$ is small enough that it does not contain any of the other image points ($1/1$, $1/3$, $1/4$, etc.). We see that

$$f^{-1}(V) = \left\{ \ldots, -\frac{5}{2}, -\frac{3}{2}, -\frac{1}{2}, \frac{1}{2}, \frac{3}{2}, \frac{5}{2}, \ldots \right\}.$$

This is not an open set. So f is not continuous.

In practice in this book, and in topology in general, when we say "Let $f : X \to Y$ be a mapping," we mean that f is a *continuous* mapping. We shall follow that custom consistently in what follows.

1.4 THE SEPARATION AXIOMS

The richness of the subject of topology begins to become evident when we examine and classify the different types of topological spaces. We do so by way of the *separation axioms*.

We begin with a sample separation axiom that is particularly intuitively appealing—just to give a flavor of this circle of ideas.

Definition 1.4.1. We say that a nonempty topological space X is a *Hausdorff space* if, for any two distinct points P, Q in X, there are open sets U and V such that

- $P \in U$ and $Q \in V$,

- $U \cap V = \emptyset$.

EXAMPLE 1.4.2. Let X be the real line with the usual topology. This is a Hausdorff space. For if P and Q are distinct points in $\mathbb{R}$ and if $\varepsilon = |P - Q|$, then the intervals $U = (P - \varepsilon/3, P + \varepsilon/3)$ and $V = (Q - \varepsilon/3, Q + \varepsilon/3)$ are neighborhoods of P and Q that are disjoint.

EXAMPLE 1.4.3. Let X be the integers with the topology that U is open if it is the complement of a finite set. Then this X is *not* a Hausdorff space. For if P, Q are distinct points, and if U, V are neighborhoods of P and Q respectively, then $U \cap V$ will always be an infinite set, and thus not empty.

The separation axioms are so important that they are numbered. Traditionally we call a Hausdorff space a T_2 space. Let us now lay out all the separation axioms. We begin with structures that are weaker than Hausdorff, and end with structures that are stronger. We shall use the terminology "neighborhood of a set S" to mean an open set that contains S.

T_0 **Space:** The nonempty space X is T_0 if, whenever $P, Q \in X$ are distinct points, then either there is a neighborhood U of P such that $Q \notin U$ or else there is a neighborhood V of Q such that $P \notin V$.

T_1 **Space:** The nonempty space X is T_1 if, whenever $P, Q \in X$ are distinct points, then there are neighborhoods U of P and V of Q such that $Q \notin U$ and $P \notin V$. It is easy to check that, in a T_1 space X, any singleton set $\{x\}$ will be closed.

T_2 **Space:** The nonempty space X is T_2 (also called *Hausdorff*) if, whenever $P, Q \in X$ are distinct points, then there are neighborhoods U of P and V of Q such that $U \cap V = \emptyset$.

T_3 **Space:** The nonempty space X is T_3 (also called *regular* if points are closed) if, whenever $P \in X$ and $F \subseteq X$ is a closed subset not containing P, then there are neighborhoods U of P and V of F so that $U \cap V = \emptyset$.

T_4 **Space:** The nonempty space X is T_4 (also called *normal* if points are closed) if, whenever E and F are disjoint closed sets in X, then there are neighborhoods U of E and V of F such that $U \cap V = \emptyset$.

Let us examine some examples that show that these separation axioms are distinct. The entire subject of point-set topology is built on examples like these. You should master them and make them part of your toolkit. It should be noted that the separation axioms (at least for spaces that are assumed to be T_1) increase in strength as the index increases. So a T_3 (regular) space is certainly T_2 (Hausdorff), and so forth.

EXAMPLE 1.4.4. Let X be the real line with the open sets being the half-lines of the form (a, ∞). It is clear that the collection $\mathcal{U}$ of such half-lines is closed under union and finite intersection. If we throw in the whole space and the empty set, then we certainly have a topology.

This space is T_0 but not T_1. To see this, note that if P and Q are distinct points of X and if $P < Q$ then $U = \{x : x > P\}$ is an open set in X. Also $Q \in U$ but $P \notin U$. So certainly X is T_0. But it is easy to see that there is no open neighborhood of P that will separate it from Q. So X is not T_1.

EXAMPLE 1.4.5. Let X be the integers equipped with the topology that U is open if its complement is finite. We have already seen that this space is not Hausdorff (Example 1.4.3). It is, however, of type T_1; because if $P, Q \in X$ are distinct then let U be the complement of $\{Q\}$ and let V be the complement of $\{P\}$. Since the space is T_1, it is also T_0.

EXAMPLE 1.4.6. Let X be the real line and equip it with the following topology. If $x \in X$ is a point other than 0, then let the neighborhoods of x be the usual intervals $U_{x,\beta} = (x - \beta, x + \beta)$ for $\beta > 0$. If $x = 0$ then let a neighborhood of x have the form

$$U_{0,\alpha} = \{t \in \mathbb{R} : -\alpha < t < \alpha, t \neq 1, 1/2, 1/3, \ldots\}$$

for $\alpha > 0$. Now generate a topology by taking all finite intersections and arbitrary unions of the described sets $U_{x,\beta}$ and $U_{0,\alpha}$. It is easy to see that the resulting space is T_2, because any two distinct points can be separated by intervals in the usual fashion—i.e., if x and y are distinct points then there are disjoint open sets U and V such that $x \in U$ and $y \in V$.

But the space is not T_3, because $E = \{1, 1/2, 1/3, \ldots\}$ is a closed set in this topology and it cannot be separated from the point 0 with open sets.

EXAMPLE 1.4.7. The *Moore plane* $\mathscr{P}$ (named after R. L. Moore (1882–1974)) is the usual closed upper halfplane $\{(x, y) : x \in \mathbb{R}, y \in \mathbb{R}, y \geq 0\}$ with the topology generated by these open sets: **(a)** If $(x, y) \in \mathscr{P}$ and $y > 0$ then consider any disc of the form $\{(s, t) \in \mathscr{P} : (s - x)^2 + (t - y)^2 < r^2\}$ for $r < y$; **(b)** If $(x, 0) \in \mathscr{P}$ then consider the singleton $\{(x, 0)\}$ union any disc of the form $\{(s, t) : (s - x)^2 + (t - r)^2 < r^2\}$ for $r > 0$. We generate the topology by taking finite intersections and arbitrary unions of the sets described in **(a)** and **(b)**. See Figure 1.5. This space is T_3 but not T_4.

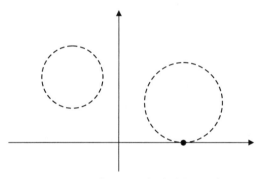

FIGURE 1.5. Open sets in the Moore plane.

To see that the space is T_3, first note that any subset of the real axis is closed. But it can easily be separated by open sets from a disjoint point in the real axis using the special open sets of type **(b)**. Points in the open upper halfplane are separated from closed sets in $\mathscr{P}$ in the usual fashion.

For non-normality, examine $E = \{(x, 0) \in \mathscr{P} : x \text{ is rational}\}$ and $F = \{(x, 0) \in \mathscr{P} : x \text{ is irrational}\}$. Then each of these sets is closed. They are clearly disjoint, but they cannot be separated by open sets. Prove these statements as an exercise.

EXAMPLE 1.4.8. Consider the interval $X = [0, 1] \subseteq \mathbb{R}$ with this topology: A set $V \subseteq [0, 1]$ is open if there is an open set $U \subseteq \mathbb{R}$ (in the usual Euclidean topology) such that $U \cap [0, 1] = V$. This X is a T_4 space. To see this let E and F be disjoint closed sets in $[0, 1]$. We claim that there is a number $\delta > 0$ such that if $e \in E$ and $f \in F$ then $|e - f| > \delta$. If not, then there are $e_j \in E$ and $f_j \in F$ with $|e_j - f_j| \to 0$. But then the common limit point x of the two sequences $\{e_j\}$ and $\{f_j\}$ would have to lie in the boundary both of E and of F. Since E and F are disjoint, that is impossible. So δ exists. Now let $U = \{x \in X : \text{dist}(x, E) < \delta/3\}$ and $V = \{x \in X : \text{dist}(x, F) < \delta/3\}$. Here $\text{dist}(x, S)$ denotes the distance of x to the set S, defined to be $\inf_{s \in S} |x - s|$. Of course $E \subseteq U$ and $F \subseteq V$. It is easy to see that U, V are both open, and they are disjoint by the triangle inequality.[1]

We refer the reader to the discussion of compactness in Section 1.5, especially Proposition 1.5.1, for further consideration of some of the ideas in the last example.

One of the most important applications of the idea of separation is the following basic result of Urysohn (commonly known as *Urysohn's lemma*—Pavel Urysohn (1898–1924)):

Theorem 1.4.1. *Let $(X, \mathcal{U})$ be a normal space and let E and F be disjoint, closed sets in X. Then there is a continuous function $f : X \to [0, 1]$ such that $f(E) = \{0\}$ and $f(F) = \{1\}$ (that is, the image of each point in E is 0 and the image of each point of F is 1).*

Remark 1.4.2. What is interesting, and significant, about Urysohn's lemma is that it relates topology to the theory of functions. In a sense, the function f in the theorem is "separating" the sets E and F.

Proof of the Theorem: By normality, there are disjoint open sets U and V such that $E \subseteq U$ and $F \subseteq V$. For technical (and also traditional) reasons we shall denote this set U by $U_{1/2}$. Now we see that E and $X \setminus U_{1/2}$ are closed and disjoint. Also $\overline{U}_{1/2}$ and F are closed and disjoint. Therefore open sets $U_{1/4}, U_{3/4}$ exist such that

$$E \subseteq U_{1/4} \ , \ \overline{U}_{1/4} \subseteq U_{1/2} \ , \ \overline{U}_{1/2} \subseteq U_{3/4} \ , \ \overline{U}_{3/4} \cap F = \emptyset.$$

Suppose now inductively that sets $U_{j/2^n}$, $j = 1, 2, \ldots, 2^n - 1$, have been defined so that

$$E \subseteq U_{1/2^n} \ , \ \ldots \ , \ \overline{U}_{(j-1)/2^n} \subseteq U_{j/2^n} \ , \ \ldots \ , \ \overline{U}_{(2^n-1)/2^n} \cap F = \emptyset.$$

Then we may continue and select sets $U_{j/2^{n+1}}$, $j = 1, \ldots, 2^{n+1} - 1$ with analogous properties.

The result of our construction is that we have, for each dyadic rational number r of the form $j/2^n$, for some $n > 0$ and $j = 1, 2, \ldots, 2^n - 1$, an open set U_r satisfying

- $E \subseteq U_r$ and $\overline{U}_r \cap F = \emptyset$,

---
[1] The reader will recall the *triangle inequality* $|a + b| \le |a| + |b|$ from calculus.

- $\overline{U}_r \subseteq U_s$ whenever $r < s$ are dyadic as above.

Now define a function $f : X \to [0, 1]$ by

$$f(x) = \begin{cases} 1 & \text{if } x \text{ belongs to no } U_r; \\ \inf\{r : x \in U_r\} & \text{if } x \text{ belongs to some } U_r. \end{cases}$$

Then $f(E) = 0$ and $f(F) = 1$. It remains to show that f is continuous. We have

Continuity at points x with $f(x) = 1$: If $x \notin \overline{U}_r$, then $f(x) \geq r$.

Continuity at points x with $f(x) = 0$: If $x \in U_r$, then $f(x) \leq r$.

Continuity at all other points: If $x \in U_s \setminus \overline{U}_r$, where $r < s$ are dyadic, then $r \leq f(x) \leq s$.

The existence of the continuous function f is now established. $\qquad \square$

Definition 1.4.9. Let $(X, \mathcal{U})$ be a topological space and $S \subseteq X$. A collection $\mathcal{W} = \{W_\alpha\}_{\alpha \in A}$ of open sets is called an *open cover* (or *open covering*) of S if

$$\bigcup_{\alpha \in A} W_\alpha \supseteq S.$$

We call an open cover $\mathcal{W}$ a *countable cover* if there are just countably many elements in the set $\mathcal{W}$. If $\mathcal{V}$ is also an open cover of S, we say that $\mathcal{V}$ is a *subcover* if each element of $\mathcal{V}$ is also an element of $\mathcal{W}$. A *countable subcover* (or *countable subcovering*) has just countably many elements. A *finite subcover* (or *finite subcovering*) has just finitely many elements.

Definition 1.4.10. Let us say that a topological space X is *Lindelöf* if every open cover of X has a countable subcover.

Proposition 1.4.3. *A regular, Lindelöf space is normal.*

Proof: Let X be as in the hypothesis, and let E, F be disjoint, closed sets in X. For each point $e \in E$, let U_e be an open set containing e such that $\overline{U}_e \cap F = \emptyset$. The set exists by the regularity hypothesis. Similarly, for each $f \in F$, we find an open set V_f such that $f \in V_f$ and $\overline{V}_f \cap E = \emptyset$. Hence E and F are Lindelöf subspaces of X. It follows that there is a countable subcover $U_{e_1}, U_{e_2}, \ldots$ of E and a countable subcover $V_{f_1}, V_{f_2}, \ldots$ of F. Now we inductively construct open sets S_j and T_j as follows:

$$\begin{aligned} S_1 &= U_{e_1} & T_1 &= V_{f_1} \setminus \overline{S}_1 \\ S_2 &= U_{e_2} \setminus \overline{T}_1 & T_2 &= V_{f_2} \setminus \overline{(S_1 \cup S_2)} \\ S_3 &= U_{e_3} \setminus \overline{(T_1 \cup T_2)} & T_3 &= V_{f_3} \setminus \overline{(S_1 \cup S_2 \cup S_3)}. \end{aligned}$$

It is now easily seen that $S \equiv \cup_j S_j$ and $T = \cup_j T_j$ are disjoint open sets containing E and F respectively. So X is normal.   □

We record now a technical result that we shall not prove, but refer the reader to [WIL, p. 57] for details.

Definition 1.4.11. Let $(X, \mathcal{U})$ be a topological space and $\mathcal{F}$ be a collection of functions from X to $\mathbb{R}$. We say that f *separates points from closed sets* if, whenever $E \subseteq X$ is a closed set and $P \notin E$, then there is an $f \in \mathcal{F}$ taking values in $I = [0, 1]$ such that $f(P) = 0$ and $f(e) = 1$ for each $e \in E$. We say that the function f *separates* P from E.

Definition 1.4.12. Let $(X, \mathcal{U})$ be a topological space. We say that X is *completely regular* if, whenever E is a closed set in X and $P \notin E$, then there is a continuous function $f : X \to I$ that separates P from E.

Definition 1.4.13. Let $(X, \mathcal{U})$ and $(Y, \mathcal{V})$ be topological spaces. An *embedding* of Y into X is a mapping $f : Y \to X$ such that

1. The mapping f is one-to-one,

2. The mapping f is continuous,

3. The mapping f^{-1} is continuous on the image of f.

We discuss embeddings in more detail in Section 1.6.

Proposition 1.4.4. *If $(X, \mathcal{U})$ is a T_1 space and $\{f_\alpha\}_{\alpha \in A}$ is a collection of functions on X (mapping to spaces X_α) that separates points from closed sets, then the evaluation mapping $e : X \to \prod_\alpha X_\alpha$ is an embedding.*

For later reference, we provide a definition and an elegant result about embedding.

Definition 1.4.14. A completely regular T_1 space will be termed a *Tychanoff space*.

Proposition 1.4.5 (Tychanoff). *Every Tychanoff space X can be embedded as a subspace of a (possibly infinite-dimensional) cube.*

Proof: Let C denote the family of all continuous real functions from X to I, the closed unit interval. The complete regularity tells us that C can distinguish points from closed sets in X. The T_1 property tells us that every singleton set is closed, hence C can also distinguish pairs of points in X. It follows that the mapping

$$f : X \to I^C$$
$$x \mapsto [f \mapsto f(x)]$$

is an embedding. (Here I^C denotes the collection of all functions from C to I.)   □

1.5 COMPACTNESS

Compact sets are fundamental in basic topology. A compact set is a possibly infinite set that behaves like a finite set. How is this possible? The concept of compact set evolved over a period of fifty or more years. It is subtle, but it is important. You should spend some time to master the idea.

Definition 1.5.1. Let X be a topological space and $K \subseteq X$. We say that K is *compact* provided that any open covering $\mathcal{W} = \{W_\alpha\}_{\alpha \in A}$ of K has a finite subcovering.

The definition of "compact" is not obvious, nor easy to understand, and requires some discussion and some examples.

EXAMPLE 1.5.2. Let X be the real numbers with the usual topology. Let $K = \{0\}$. So K is a set with a single point (a *singleton*). Then K is compact.

If $\mathcal{W} = \{W_\alpha\}_{\alpha \in A}$ is an open covering of K then one of the W_α, say W_{α_1}, contains 0. Then the finite subcovering $\{W_{\alpha_1}\}$ will do the job.

EXAMPLE 1.5.3. Let X be the real numbers with the usual topology. Let S be the integers $\mathbb{Z}$. Then S is not compact. For let $\mathcal{W}$ be the open cover consisting of the intervals $W_k = (k - 2/3, k + 2/3)$ for the indices $k = \cdots -3, -2, -1, 0, 1, 2, 3, \ldots$. Certainly $\mathcal{W}$ is a covering of the set S. But each integer k lies just in W_k and in no other element of the cover. So there is no subcovering that will still cover S. In particular, there is no finite subcovering. Thus S is not compact.

EXAMPLE 1.5.4. Let X be the real numbers with the usual topology. Let

$$K = \left\{ 1, \frac{1}{2}, \frac{1}{3}, \frac{1}{4}, \ldots \right\} \bigcup \{0\}.$$

Then K is compact. To see this, let $\mathcal{W} = \{W_\alpha\}$ be an open covering of K. Then some W_{α_0} contains 0. Note that W_{α_0} is an open set in $\mathbb{R}$, so is a union of open intervals. But then W_{α_0} will contain all $1/j$ for $j > J_0$, some J_0 sufficiently large. See Figure 1.6. Now select W_{α_1} that contains $1/1$, W_{α_2} that contains $1/2$, on up to $W_{\alpha_{J_0}}$ that contains $1/J_0$. We may conclude that the open sets

$$W_{\alpha_0}, W_{\alpha_1}, W_{\alpha_2}, \ldots, W_{J_0}$$

form a finite subcover of the original covering $\mathcal{W}$. Thus K is compact.

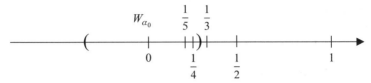

FIGURE 1.6. Compactness of the set K.

EXAMPLE 1.5.5. Let X be the real numbers with the usual topology. Let $K = [0, 1]$. Then K is compact. This assertion is not so easy to prove, though you will recognize elements of Example 1.5.4 in the argument that we are about to present.

Fix an open cover $\mathcal{W} = \{W_\alpha\}$ of $K = [0, 1]$. Let

$$S = \{b \in [0, 1] : \text{the interval } [0, b] \text{ has a finite subcover}\}.$$

Then S is not empty since $0 \in S$; that is to say, the singleton $\{0\}$ has a finite subcover (again see Example 1.5.2). Also S is bounded above—by 1! Let s_0 be the supremum[2] of S. Of course $s_0 \in [0, 1]$. Seeking a contradiction, we suppose that $s_0 < 1$.

Let W_{α_0} be a member of the open covering that contains s_0. Look at Figure 1.7. Since s_0 is the supremum of S, there are elements of S (to the left of s_0) that are arbitrarily close to s_0. Choose one such that lies inside W_{α_0}. Call it s. The interval $[0, s]$ has a finite subcovering $W_{\alpha_1}, \ldots, W_{\alpha_k}$ by the definition of S. But then $[0, s_0]$ has the finite subcovering $W_{\alpha_0}, W_{\alpha_1}, \ldots, W_{\alpha_k}$. In fact there are points s' to the right of s_0 that lie in W_{α_0}, and we may note that the interval $[0, s']$ has *the very same* finite subcover. But this contradicts the choice of s_0 as the supremum of S, which implies that $s_0 = 1$ and the entire interval $[0, 1]$ has a finite subcover.

FIGURE 1.7. Compactness of the unit interval.

The last example illustrates the important point that it is not always easy to see that even a simple set like the unit interval is compact. We need some theorems to help us in the process of identifying compact sets.

---

[2]It is a fundamental idea of real analysis that any bounded set $S \subseteq \mathbb{R}$ has a supremum (and an infimum). The supremum is the least real number M such that no element of S is greater than M. The infimum is the greatest real number m such that no element of S is less than m.

Having looked at some concrete examples, we now begin to assemble some ideas that will help us to understand compact sets.

Proposition 1.5.1. *Let K be a compact set in a Hausdorff (i.e., T_2) space and let x be a point that is not in K. Then there are disjoint open sets U and V such that $U \supseteq K$ and $V \ni x$.*

Proof: This proof is a nice illustration of how compactness works. By the Hausdorff property, for each point k in K, there is a neighborhood U_k of k and a neighborhood V_k of x such that $U_k \cap V_x = \emptyset$. The sets $\{U_k\}$ form an open cover of K. So there is a finite subcover $\{U_{k_1}, U_{k_2}, \ldots, U_{k_m}\}$ of K. But then $\mathcal{V} \equiv V_{k_1} \cap V_{k_2} \cap \cdots \cap V_{k_m}$ is an open neighborhood of x and $\mathcal{U} \equiv U_{k_1} \cup U_{k_2} \cup \cdots \cup U_{k_m}$ is an open neighborhood of K. The sets $\mathcal{U}$ and $\mathcal{V}$ are disjoint and separate K and x.   □

Remark 1.5.2. In the last proof we used the word "separate" in a technical, mathematical sense. It meant that $\mathcal{U} \supseteq K$, $\mathcal{V} \ni x$, and $\mathcal{U} \cap \mathcal{V} = \emptyset$. We shall use the word "separate" in an analogous fashion throughout the book.

Proposition 1.5.3. *A compact set in a Hausdorff space is closed.*

Proof: Let K be a compact set and x a point that is not in K. By the preceding proposition, there is a neighborhood U of x that is disjoint from K. That shows that the complement of K is open. So K is closed.   □

Proposition 1.5.4. *Let $f : X \to Y$ be a continuous mapping of topological spaces. If $K \subseteq X$ is compact then $f(K) \equiv \{f(k) : k \in K\}$ is compact.*

Proof: Let $\mathcal{W} = \{W_\alpha\}_{\alpha \in A}$ be an open covering of $f(K)$. Then, since f is continuous, $\{f^{-1}(W_\alpha)\}_{\alpha \in A}$ is an open covering of K. Therefore there is a finite subcovering $f^{-1}(W_{\alpha_1}), f^{-1}(W_{\alpha_2}), \ldots, f^{-1}(W_{\alpha_m})$. It follows then that $W_{\alpha_1}, W_{\alpha_2}, \ldots, W_{\alpha_m}$ is a finite subcovering of $f(K)$. Thus $f(K)$ is compact.   □

Theorem 1.5.5 (Heine-Borel). *A set $E \subseteq \mathbb{R}$ is compact if and only if it is closed and bounded.*

Proof: If the set is closed and bounded then compactness follows precisely as in the proof of Example 1.5.5. We leave the details to the reader.

Now suppose that $E \subseteq \mathbb{R}$ is compact. Since $\mathbb{R}$ is Hausdorff, we can be sure by Proposition 1.5.3 that E is closed. It remains to show that E is bounded. If E is not bounded, then let $E \ni e_j \to +\infty$ (the case $e_j \to -\infty$ is handled in exactly the same fashion). We may assume, passing to a subsequence if necessary, that $|e_j - e_{j+1}| \geq j$. Let $\mathcal{W} = \{W_\alpha\}$ be a covering of E such that no W_α has diameter greater than $1/2$. It follows that any subcovering will have to have distinct elements that cover e_1, e_2, etc. So there must be infinitely many elements in any subcovering of the set E, hence there is no finite subcovering.   □

The Heine-Borel theorem is an extremely useful result, and makes it straightforward to check whether any set in $\mathbb{R}$ is compact. The result also holds in $\mathbb{R}^N$ for any N, and we shall say more about that matter later.

Proposition 1.5.6. *Let X be a topological space. A closed subset of a compact set in X is compact.*

Proof: Let K be a compact set and $E \subseteq K$ a closed subset. Let $\mathcal{W} = \{W_\alpha\}_{\alpha \in A}$ be an open covering of E. Let $W' = X \setminus E$. Then W' is open and $\mathcal{W} \cup \{W'\}$ is an open covering of K. Hence there is a finite subcovering of K given by

$$W', W_{\alpha_1}, W_{\alpha_2}, \ldots, W_{\alpha_m} .$$

But then

$$W_{\alpha_1}, W_{\alpha_2}, \ldots, W_{\alpha_m}$$

is a finite subcover of E itself. Thus E is compact.   □

Proposition 1.5.7. *A one-to-one, continuous map from a compact space X onto a Hausdorff space Y is bicontinuous.*

Proof: Let f be such a map. Let U be an open subset of X. Then $E = X \setminus U$ is closed. Hence, by 1.5.6, it is compact. It follows from 1.5.4 that $f(E)$ is compact in Y. So, by 1.5.3, $f(E)$ is closed. But then $f(U) = Y \setminus f(E)$ is open. So f is an open mapping. But that says that f^{-1} is continuous. □

We next turn to one of the more profound and useful results about compactness. It is necessary to begin with a definition.

Definition 1.5.6. Let X be a topological space. Let $\mathcal{F} = \{F_\alpha\}_{\alpha \in A}$ be a family of sets in X. We say that $\mathcal{F}$ has the *finite intersection property* if, whenever F_{α_1}, F_{α_2}, ..., F_{α_m} is a finite collection of elements of $\mathcal{F}$, then $\cap_{j=1}^m F_{\alpha_j}$ is nonempty.

Theorem 1.5.8. *A topological space $(X, \mathcal{U})$ is compact if and only if any family $\mathcal{F} = \{F_\alpha\}_{\alpha \in A}$ of closed sets in X with the finite intersection property actually satisfies $\cap_{\alpha \in A} F_\alpha \neq \emptyset$.*

Proof: First suppose that X is compact. Let $\mathcal{F} = \{F_\alpha\}_{\alpha \in A}$ be a family of closed sets in X and suppose that $\cap_{\alpha \in A} F_\alpha = \emptyset$. Now look at $\{X \setminus F_\alpha\}$. This must (by De Morgan's law) then be an open cover of X. Since X is compact, there is a finite subcover $X \setminus F_{\alpha_1}, X \setminus F_{\alpha_2}, \ldots, X \setminus F_{\alpha_m}$. But this says (again by De Morgan's law) that $F_{\alpha_1} \cap F_{\alpha_2} \cap \cdots \cap F_{\alpha_m} = \emptyset$. So $\mathcal{F}$ does not have the finite intersection property. That proves one direction of the theorem.

Now suppose that, whenever $\mathcal{F} = \{F_\alpha\}_{\alpha \in A}$ is a family of closed sets with the finite intersection property, then $\cap_{\alpha \in A} F_\alpha \neq \emptyset$. Our job then is to show that X is compact. Let $\mathcal{U} = \{U_\alpha\}_{\alpha \in A}$ be an open cover of X. Now consider the family $\mathcal{F} \equiv \{X \setminus U_\alpha : \alpha \in A\}$. Then $\mathcal{F}$ is a family of closed sets and, by De Morgan's law, $\cap_\alpha X \setminus U_\alpha = \emptyset$. So there must be finitely many $X \setminus U_{\alpha_1}, X \setminus U_{\alpha_2}, \ldots, X \setminus U_{\alpha_m}$ with empty intersection. But, again by De Morgan's law, this says that $U_{\alpha_1}, U_{\alpha_2}, \ldots, U_{\alpha_m}$ is a finite subcover of the family $\mathcal{U}$. Thus X is compact. $\qquad\square$

We close this section with a notion that will come up later in the book.

Definition 1.5.7. Let $(X, \mathcal{U})$ be a topological space. Let $x \in X$. We say that a collection $\mathcal{W}$ of neighborhoods of x is a *neighborhood base* (or neighborhood basis) at x if every neighborhood of x contains an element of $\mathcal{W}$.

Clearly "neighborhood base" is a local version of the idea of topology.

Definition 1.5.8. Let $(X, \mathcal{U})$ be a topological space. We say that X is *locally compact* if each point of X has a neighborhood base consisting of sets whose closures are compact (such sets are often called *precompact*).

EXAMPLE 1.5.9. Let X be the real numbers with the usual topology. Let $x \in X$. Then the sets $(x - \varepsilon, x + \varepsilon)$ for $\varepsilon > 0$ form a neighborhood base for the point x, and each of these sets has compact closure. So X is locally compact.

1.6   HOMEOMORPHISMS

The most fundamental tool in the subject of point-set topology is the homeomorphism. This is how we measure the equivalence or inequivalence of topological spaces. This section will introduce you to the idea, and provide several examples.

Definition 1.6.1. Let X and Y be topological spaces. A mapping $f : X \to Y$ is said to be a *homeomorphism* if

- The mapping f is one-to-one and onto,

- The mapping f is continuous,

- The mapping f^{-1} is continuous.

If the mapping f is everything but onto then we call it an *embedding*.

It is plain that a homeomorphism f preserves open sets, closed sets, and compact sets. So does f^{-1}. Thus all the essential features of a topology are transferred naturally under a homeomorphism. If $f : X \to Y$ is a homeomorphism then we say that X and Y are *homeomorphic*.

EXAMPLE 1.6.2. The set $S = \{(x, y) \in \mathbb{R}^2 : x^2 + y^2 = 1\}$ and the set $T = \{(x, y) \in \mathbb{R}^2 : 4x^2 + y^2 = 1\}$ are homeomorphic. The mapping $f : S \to T$ given by $(x, y) \mapsto (x/2, y)$ is the needed homeomorphism. See Figure 1.8.

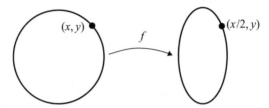

FIGURE 1.8. Homeomorphism of the circle and the ellipse.

Example 1.6.2 is a simple example, but it captures the spirit of what a homeomorphism is. For look at Figure 1.8. We see that the circle S and the ellipse T have different shapes in the heuristic sense. But there is an essential sameness to them. You can get one from the other by a continuous deformation—a *bending and stretching without tearing*.[3] That is what the concept of homeomorphism is all about.

EXAMPLE 1.6.3. Let

$$S = \{(x, y) \in \mathbb{R}^2 : 0 \le x \le 1, y = 0\}$$

and

$$T = \{(x, y) \in \mathbb{R}^2 : 0 \le x \le 1/2, y = x\}$$
$$\cup \{(x, y) \in \mathbb{R}^2 : 1/2 \le x \le 1, y = 1 - x\}.$$

Then S and T are homeomorphic. See Figure 1.9.

---

[3]This is a good heuristic for what a homeomorphism is, but it is not completely accurate. For example, a Möbius band with a single half-twist and Möbius band with three half-twists are homeomorphic. But bending and stretching do not provide a complete explanation of why that is so.

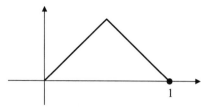

FIGURE 1.9. Homeomorphism of a segment with a bent, piecewise linear curve.

In fact the homeomorphism is

$$f : S \to T$$
$$(x, 0) \mapsto \begin{cases} (x, x) & \text{if } 0 \le x \le 1/2 \\ (x, 1 - x) & \text{if } 1/2 < x \le 1. \end{cases}$$

We leave it to the reader to verify the details that this is indeed a homeomorphism. Once again we see that a homeomorphism represents a bending and stretching (without tearing).

EXAMPLE 1.6.4. Let $S = \{(x, y) \in \mathbb{R}^2 : |x| \le 1, |y| \le 1\}$ and $T = \{(x, y) \in \mathbb{R}^2 : x^2 + y^2 \le 1\}$. Then S and T are homeomorphic.

To see this, assign to each point $(x, y) \in S$ a number $t = t(x, y)$, $0 \le t \le 1$ that represents the fraction of the distance that the point (x, y) is from the origin to the boundary. See Figure 1.10. Now define

$$f : S \to T$$
$$(x, y) \mapsto \frac{t(x, y)}{\sqrt{x^2 + y^2}} \cdot (x, y),$$

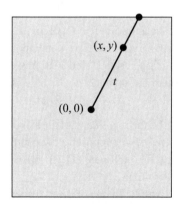

FIGURE 1.10. The function that represents fraction of the distance to the boundary.

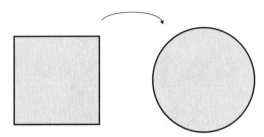

FIGURE 1.11. A homeomorphism of the square to the disc.

where it is understood that $f(0, 0) = (0, 0)$. It is easy to check that f maps S to T in a one-to-one, onto, bicontinuous fashion. See Figure 1.11.

The next result may seem a bit surprising.

EXAMPLE 1.6.5. The spaces $X = (0, 1) \subseteq \mathbb{R}$ and $Y = (0, \infty)$ are homeomorphic. For let $f : X \to Y$ be given by $f(x) = x/(1 - x)$. Then it is easy to verify that f satisfies all the properties of a homeomorphism of these spaces.

This last example illustrates the idea that a homeomorphism does not preserve size or distance. Instead, it preserves the essential topological nature, or the shape, of a space.

1.7 CONNECTEDNESS

Certainly we have an intuitive idea of what it means for a space to be connected: A space is connected if it is "all of one piece"; the space is disconnected if it is in "several pieces". Now we shall make these ideas precise.

Definition 1.7.1. Let X be a topological space and $E \subseteq X$ a nonempty set. We say that E is *disconnected* if there are nonempty $F, G \subseteq X$ such that $E = F \cup G$ and there are open sets U and V in X so that $F \subseteq U, G \subseteq V$, and $U \cap V = \emptyset$. If there are no such E, F, U, V then we say that E is *connected*.

EXAMPLE 1.7.2. Let X be the real line with the standard topology and $I = [0, 1]$. Then I is connected. For suppose to the contrary that $I = F \cup G$, F and G both nonempty, $F \subseteq U$ open, $G \subseteq V$ open, and $U \cap V = \emptyset$. We derive a contradiction as follows.

Certainly the point 1 will lie in one of the two open sets; say that $1 \in U$. Let $c = \sup_{v \in V} v$. Since there is a neighborhood of 1 that lies entirely in U, we can be sure that c is not in U. So $c \in V$. Thus c has a neighborhood that lies entirely in V. But, since c is the supremum of V, c has points to the immediate left that lie in V. Also c has points to the immediate right that lie in U. Hence c does not have a neighborhood that lies entirely in V. In conclusion, U and V do not exist and the interval I is connected.

Proposition 1.7.1. *Let $f : X \to Y$ be a continuous mapping. Let $E \subseteq X$ be connected. Then $f(E)$ is connected.*

Proof: Suppose to the contrary that $f(E)$ is disconnected. Write $f(E) = A \cup B$ (both nonempty) with disjoint open sets U and V so that $A \subseteq U$ and $B \subseteq V$. Then $f^{-1}(U)$ and $f^{-1}(V)$ are disjoint, nonempty open sets in X that separate E, meaning that E is disconnected. This of course is a contradiction. So $f(E)$ must be connected.   □

EXAMPLE 1.7.3. Consider the topology on the real line generated by intervals of the form $[a, b)$ or $[a, +\infty)$ (here we mean "generated" in the sense of taking finite intersection and arbitrary union). This is the Sorgenfrey line, which we encountered first in Example 1.2.5. The Sorgenfrey line is one of the most important examples in topology. We show here that the Sorgenfrey line is disconnected.

If (c, d) is any open interval then

$$(c, d) = \bigcup_{\varepsilon > 0} [c + \varepsilon, d).$$

Thus (c, d) is the union of Sorgenfrey open sets. So any standard open interval is open in the Sorgenfrey topology. We see, then, that the Sorgenfrey topology contains all the usual open sets and some new ones as well.

We conclude then that $(-\infty, 0)$ and $[0, \infty)$ are both open. And certainly $\mathbb{R} = (-\infty, 0) \cup [0, \infty)$. So $\mathbb{R}$ is disconnected in the Sorgenfrey topology.

EXAMPLE 1.7.4. The *topologist's sine curve* (another famous example) is the set

$$S = \{(0, y) : y \in \mathbb{R}, -1 \le y \le 1\} \bigcup \left\{ \left(x, \sin \frac{1}{x}\right) : x > 0 \right\} .$$

See Figure 1.12.

It is connected. For certainly the left-hand portion of S, which is $\{(0, y) : |y| \le 1\}$, is connected. And any open set that contains that portion will contain a neighborhood of the origin and hence intersect the right-hand portion (which gets arbitrarily close to the origin).

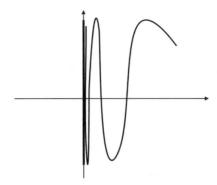

FIGURE 1.12. The topologist's sine curve.

Proposition 1.7.2. *Let* $(X, \mathcal{U})$ *be a topological space. If A and B in X are connected sets with a common point p then* $A \cup B$ *is connected.*

Proof: Suppose not. Say that the disjoint open sets U and V disconnect $A \cup B$. Then p must lie in one of these two open sets. Say that it lies in U. Since A cannot be disconnected, it follows that $A \subseteq U$. A similar argument shows that $B \subseteq U$. Thus $A \cup B \subseteq U$ and it is not the case that U and V disconnect $A \cup B$. This contradiction shows that $A \cup B$ is connected.   □

We say that two points x and y in a topological space X are *connected in X* if there is a connected set $S \subseteq X$ that contains both x and y. It is easy to verify that this is an equivalence relation. The resulting equivalence classes are called *connected components.*

Of course a connected topological space X has just one connected component. As a counterpoint, the set

$$X = \{(x, y) \in \mathbb{R}^2 : |(x, y) \pm (4, 0)| = 1\}$$

(where we are using the standard Euclidean topology in the plane, as in Example 1.2.2) has two connected components. See Figure 1.13.

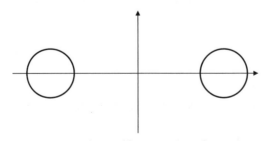

FIGURE 1.13. A set with two connected components.

1.8   PATH-CONNECTEDNESS

There is a stronger concept of connectedness that is particularly useful in
Euclidean space. We treat it in this section.

Definition 1.8.1. Let $(X, \mathcal{U})$ be a topological space. We say that X is *path-
connected* if, given any two points $P, Q \in X$, there is a continuous path
(i.e., a *function*) $\gamma : [0, 1] \to X$ such that $\gamma(0) = P$ and $\gamma(1) = Q$.

EXAMPLE 1.8.2. Let X be the open unit disc in the Euclidean plane. Then
X is path connected. For if $P, Q \in X$ then the path

$$\gamma(t) = (1 - t)P + tQ$$

will connect the two points.

EXAMPLE 1.8.3. Consider the topologist's sine curve as in Example 1.7.4:

$$S = \{(0, y) : y \in \mathbb{R}, -1 \le y \le 1\} \bigcup \left\{ \left(x, \sin \frac{1}{x} \right) : x > 0 \right\} .$$

We know that this set is connected. But it is not path-connected. The point
$(0, 0)$ lies in S as does the point $(2/\pi, 1)$. Suppose that γ is a continuous
path connecting them. We may take it that $\gamma(0) = (0, 0)$. But then there
are points t arbitrarily close to 0, of the form $2/[(2k + 1)\pi]$, at which the
function $\sin \frac{1}{x}$ takes the values ± 1. Thus $\lim_{t \to 0} \gamma(t) \ne \gamma(0)$, and γ cannot
be continuous (remember the ε-δ definition of continuity from calculus—
we proved it equivalent to the "inverse image of open sets" definition in
Proposition 1.3.3). So γ cannot be continuous.

Proposition 1.8.1. Let $(X, \mathcal{U})$ be a topological space. If X is path-connected
then X is connected.

Proof: Suppose to the contrary that X is disconnected. So there are disjoint
open sets U, V that disconnect X. Let P be a point of $U \cap X$ and Q be a
point of $V \cap X$ and $\gamma : [0, 1] \to X$ a path that connects them. Then $\gamma^{-1}(U)$
and $\gamma^{-1}(V)$ disconnect the unit interval $[0, 1]$, and that is impossible.   □

Proposition 1.8.2. Every connected open set U in Euclidean space is path-
connected.

Proof: Fix a point $P \in U$. Define

$$S = \{u \in U : \text{the point } u \text{ can be connected to } P \text{ by a path}\} .$$

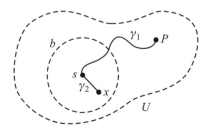

FIGURE 1.14. Connected open sets in Euclidean space are path-connected.

Then S is nonempty because P itself is in the set. Also, S is open because if $s \in S$ then $s \in U$, so there is a small ball **b** around s that lies in U. We know that there is a path γ_1 that connects P to s. If $x \in \mathbf{b}$ then there is certainly a linear path γ_2 that connects s to x. The concatenation of γ_1 and γ_2 connects P to x. Refer to Figure 1.14.

Hence the ball **b** lies in S and S is open. Finally, if $t \in U$ and $t \notin S$ then t cannot be connected to the point P by a path. But of course $t \in U$, so there is a small ball $\mathbf{b}'$ about t that lies in U, and points of $\mathbf{b}'$ can be connected to t by a linear path. Thus they cannot be connected to P. So the entire ball is in the complement of S. We conclude that the complement of S is open hence S is closed.

Since S is nonempty, open, and closed, we conclude by the connectedness of Euclidean space that S is all of U. That gives the desired conclusion.

□

1.9 CONTINUA

The heuristic model for a continuum is (the image of) a curve in the plane. A continuum is a set that should have no cuts or breaks. A formal definition is as follows.

Definition 1.9.1. Let X be a compact, connected Hausdorff space. Then X is a *continuum*.

EXAMPLE 1.9.2. The unit interval $I = [0, 1]$, the unit circle in the plane, and the torus are all continua.

Our goal in this discussion is to develop useful topological characterizations of continua.

Definition 1.9.3. Let X be a connected T_1 space. A *cut point* of X is a point $p \in X$ such that $X \setminus \{p\}$ is disconnected. If p is *not* a cut point of X, then we say that p is a *noncut point*. A *cutting* of X is a triple (p, U, V) where p is a cut point for X and the open sets U, V separate $X \setminus \{p\}$.

Lemma 1.9.1. *If K is a metric space that is a continuum, and if K has exactly two noncut points, then K is homeomorphic to the unit interval.*

Proof: Although this result is intuitively appealing, it is remarkably tricky to prove (relying as it does on the construction of the real numbers and other subtle ideas). We refer the reader to [WIL, pp. 206–207] for the details. $\square$

In what follows, a metric space is a set X with a notion d of distance. We assume that the metric $d(x, y)$ is non-negative, that d is symmetric in its arguments, that $d(x, y) = 0$ if and only if $x = y$, and that d satisfies a triangle inequality $d(x, y) \le d(x, z) + d(z, y)$. We shall discuss metric spaces in more detail in Section 1.12.

Theorem 1.9.2. *Let K be a metric space and assume that K is a continuum. Further suppose that, for any two distinct points $p, q \in K$, the set $K \setminus \{p, q\}$ is disconnected. Then K is homeomorphic to the unit circle.*

The theorem shows that the property of having cuts induces an order on the topological space in question. It is a remarkable characterization of the circle.

Proof of Theorem 1.9.2: First let us show that K has no cut points. Suppose to the contrary that (p, U_0, V_0) is a cutting. Then, since $U_0 \cup \{p\}$ and $V_0 \cup \{p\}$ are both continua, each contains at least some noncut points. Say that y is a noncut point of $U_0 \cup \{p\}$ and z is a noncut point of $V_0 \cup \{p\}$. Then the connected sets $(U_0 \cup \{p\}) \setminus \{y\}$ and $(V_0 \cup \{p\}) \setminus \{z\}$ intersect; and by 1.7.2 their union $K \setminus \{y, z\}$ is therefore connected. This contradicts our hypothesis. So K has no cut points.

Now, following the statement of the theorem, let p and q be arbitrary distinct points of K. Then $K \setminus \{p, q\} = U \cup V$, where U and V are nonempty, disjoint, open subsets of K. Let $U^* = U \cup \{p, q\}$ and $V^* = V \cup \{p, q\}$. We claim that U^*, V^* are arcs with p, q as endpoints. Moreover, $U^* \cap V^* = \{p, q\}$. This shows that $K = U^* \cup V^*$ is homeomorphic to a circle.

Certainly U^* and V^* are both connected. To see this, suppose that $U^* = S \cup T$, with S, T open, nonempty, and disjoint in U^*. If S contains both p and q, then T is open in U and hence in K. This is impossible, since T is closed in U^* hence closed in K (and K is connected). As a result, we may suppose that $p \in S$ and $q \in T$. But now the same reasoning shows that $S \setminus \{p\}$ is open and closed in the connected set $K \setminus \{p\}$. That is impossible, so U^* and V^* are connected.

Next we assert that p and q are both noncut points of U^* (and also of V^*). For if S and T disconnect $U^* \setminus \{p\}$, and if $q \in S$, then (again by the arguments that we presented above), T is both open and closed in $K \setminus \{p\}$. That is impossible.

Finally, we wish to show that U^* and V^* each has precisely two noncut points, namely p and q. We proceed as follows:

- Say that each of U^* and V^* has a third noncut point. Say that p' is a noncut point of U^* and q' is a noncut point of V^* that are distinct from p and q. Then the sets $U^* \setminus \{p'\}$ and $V^* \setminus \{q'\}$ are connected, they intersect, and their union is $K \setminus \{p', q'\}$ (which is a disconnected set). This is a contradiction, so these third noncut points do not exist.

- Suppose that just one of our two sets, say U^*, has a third noncut point p''. Then, if q'' is any point in V, we have a cutting (q'', L, M) of V^*, where L and M are connected and $p \in L$, $q \in M$. (Clearly not both p and q can belong to the same one of these sets.) Now $U^* \setminus \{p'\}$, L, and M form a chain of connected sets whose union is $K \setminus \{y, z\}$, a contradiction.

So we see that each of U^* and V^* is a metric continuum with precisely two noncut points, p and q, and $U^* \cap V^* = \{p, q\}$. In fact we know that each of U^* and V^* is an arc with p, q as endpoints. It follows that $K = U^* \cup V^*$ is homeomorphic to the circle.   □

1.10   TOTALLY DISCONNECTED SPACES

Of course a connected topological space has just one connected component. A totally disconnected space is just the opposite. We explore that new idea here.

Definition 1.10.1. A topological space X is *totally disconnected* if the connected components of X are single points.

EXAMPLE 1.10.2. Let X be the rational numbers $\mathbb{Q}$. Then X is totally disconnected. For certainly any subset of X that contains at least two points is disconnected.

Similarly, if X is the irrational numbers $\mathbb{R} \setminus \mathbb{Q}$, then X is totally disconnected.

EXAMPLE 1.10.3. Refer ahead to the construction of the Cantor set $\mathbf{C}$ in the next section. We now think of the Cantor set as lying in the real line which is in turn the x-axis of the coordinate plane—see Figure 1.15.

Let Q be the set of endpoints of the intervals that are deleted from $I = [0, 1]$ in order to construct $\mathbf{C}$. Set $P = \mathbf{C} \setminus Q$. Fix the point $p = (1/2, 1/2)$

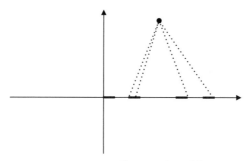

FIGURE 1.15. Construction of L_x.

in the plane. Of course the Cantor set is a subset of the x-axis. If $x \in \mathbf{C}$ then let L_x be the line segment joining p to x. Define

$$L_x^* = \{(x_1, x_2) \in L_x : x_2 \text{ is rational}, x_1 \in Q\}$$
$$\cup\{(x_1, x_2) \in L_x : x_2 \text{ is irrational}, x_1 \in P\}.$$

See Figure 1.15.

Now set

$$\mathbf{K} = \bigcup_{x \in \mathbf{C}} L_x^*.$$

It is plain that $\mathbf{K}$ is connected (in the usual topology of the plane), while $\mathbf{K} \setminus \{p\}$ is totally disconnected.

If $(X, \mathcal{U})$ and $(Y, \mathcal{V})$ are topological spaces, then we can equip the set-theoretic product $X \times Y \equiv \{(x, y) : x \in X, y \in Y\}$ with a topology as follows: We generate a topology with all sets of the form $U \times V$, where $U \in \mathcal{U}$ and $V \in \mathcal{V}$, by taking all possible finite intersections and arbitrary unions. This "product topology" will be considered in more detail in Section 2.2.

Proposition 1.10.1. *The product of totally disconnected spaces is totally disconnected. Also every subspace of a totally disconnected space is totally disconnected.*

Proof: Let X_1 and X_2 be totally disconnected and let π_j be the projection from $X = X_1 \times X_2$ to X_j: $\pi_j(x_1, x_2) = x_j$. Of course the projection is a continuous mapping. If S is a nonempty, connected subset of X then $\pi_j(S)$ is a connected subset of X_j. So $\pi_j(S)$ is a point, $j = 1, 2$. It follows then that S is a point.

The second statement of the proposition is nearly obvious, and we leave it as an exercise.   □

Definition 1.10.4. Let $(X, \mathcal{U})$ be a topological space. A subset $S \subseteq X$ is said to be *dense* in X if, for each $x \in X$ and each neighborhood U of x there is a point $s \in S$ such that $s \in U$.

EXAMPLE 1.10.5. Let X be the real numbers with the usual topology. Then the rational numbers $\mathbb{Q}$ are dense in X.

In dimension theory, we define topological dimension inductively. Let us say that a space $(X, \mathcal{U})$ is *separable* if it has a countable dense subset. Suppose that $(X, \mathcal{U})$ is a separable metric space. We begin by positing that the empty set (and *only* the empty set) has dimension -1. Next, X has dimension $\leq N$ if each point of X has a neighborhood basis of open sets U such that ∂U has dimension (in the space X) that is $\leq (N - 1)$.

EXAMPLE 1.10.6. The set

$$X = \{1, 1/2, 1/3, \ldots\} \cup \{0\}$$

in $\mathbb{R}$ has dimension 0. The proof is obvious. Any of the points $1/j$ has a neighborhood basis $(1/j - 1/4^j, 1/j + 1/4^j) \cap X$ with boundaries that are empty—hence they are of dimension -1. Also the point 0 has a neighborhood basis $(0 - \pi/10^j, 0 + \pi/10^j)$ with boundaries that are empty—hence they are of dimension -1. We conclude that X has dimension ≤ 0. But the set obviously is not of dimension -1. So it has dimension 0.

EXAMPLE 1.10.7. The set

$$X = \{(x, y) \in \mathbb{R}^2 : x^2 + y^2 < 1\}$$

has dimension 2. We calculate as follows: If $(x, y) \in X$, then let $\delta = \sqrt{1 - x^2 - y^2}$. Then $U_j = \{(s, t) : |(x, y) - (s, t)| < 1/j\}$ for $j > 2/\delta$ is a neighborhood system with $\partial U_j \cap X$ having dimension 1 (that is a separate, but plausible, calculation). Hence the dimension of X is ≤ 2. But it is easy to check that the dimension of X is not ≤ 1. So the dimension of X is precisely 2.

Lemma 1.10.2. *A nonempty subset S of a 0-dimensional space X is 0-dimensional.*

Proof: The set S clearly has dimension at most 0. It is not empty, so does not have dimension -1. So the dimension must be 0.   □

Proposition 1.10.3. *A nonempty space X has dimension 0 if and only if every point $p \in X$ and every closed set $C \subseteq X$ such that $p \notin C$ can be separated by open sets.*

Proof: First suppose that X has dimension 0 according to our original definition. Certainly $X \setminus C$ is a neighborhood of p. So there is a set V with $p \in V \subseteq X \setminus C$ and V is both open and closed. But $V \cap C = \emptyset$, so p and C are separated.

The converse is proved similarly.   □

Lemma 1.10.4. *A connected, 0-dimensional space consists of just one point.*

Proof: Suppose instead that the space contains two points. Then Proposition 1.10.3 tells us that these points are separated. So the space is in fact disconnected, and that is a contradiction.   □

Proposition 1.10.5. *A 0-dimensional space is totally disconnected.*

Proof: This follows from 1.10.2 and 1.10.3.   □

1.11 THE CANTOR SET

The Cantor set is one of the most remarkable constructions in all of mathematics. It is found in many areas of analysis, topology, and geometry, and frequently arises as an example or counterexample in various theories. We describe it here, and derive some of its properties. The set only begins to suggest the richness of the structure of the real number system.

We begin with the unit interval $S_0 = [0, 1]$. We extract from S_0 its open middle third; thus $S_1 = S_0 \setminus (1/3, 2/3)$ (Figure 1.16), which consists of two closed intervals of equal length $1/3$.

$$0 \hspace{10cm} 1$$

FIGURE 1.16. First step in the construction of the Cantor set.

Now we construct S_2 from S_1 by extracting from each of its two intervals the middle third: $S_2 = [0, 1/9] \cup [2/9, 3/9] \cup [6/9, 7/9] \cup [8/9, 1]$. Figure 1.17 shows S_2.

$$0 \hspace{10cm} 1$$

FIGURE 1.17. Second step in the construction of the Cantor set.

Continuing in this fashion, we construct S_{j+1} from S_j by extracting the middle third from each of its component subintervals. We define the Cantor set C to be

$$C = \bigcap_{j=1}^{\infty} S_j \, .$$

Each of the sets S_j is closed and bounded, hence compact. By Theorem 1.5.8, C is therefore not empty. The set C is closed and bounded, hence compact.

Proposition 1.11.1. *The Cantor set C has zero length, in the sense that the complementary set $[0, 1] \setminus C$ has length 1.*

Proof: In the construction of S_1, we removed from the unit interval one interval of length 3^{-1}. In constructing S_2, we further removed two intervals of length 3^{-2}. In constructing S_j, we removed 2^{j-1} intervals of length 3^{-j}. Thus the total length of the intervals removed from the unit interval is

$$\sum_{j=1}^{\infty} 2^{j-1} \cdot 3^{-j} = \frac{1}{3} \cdot \sum_{j=0}^{\infty} \left(\frac{1}{3}\right)^j .$$

The geometric series sums easily and we find that the total length of the intervals removed is

$$\frac{1}{3}\left(\frac{1}{1 - 2/3}\right) = 1 .$$

Thus the Cantor set has length zero because its complement in the unit interval has length one. □

Proposition 1.11.2. *The Cantor set is uncountable.*

Proof: We assign to each element of the Cantor set a label consisting of a sequence of 0s and 1s that identifies its location in the set.

Fix an element x in the Cantor set. Then certainly x is in S_1. If x is in the left half of S_1, then the first digit in the label of x is 0; otherwise it is 1. Likewise $x \in S_2$. By the first part of this argument, it is either in the left half S_{21} of S_2 (when the first digit in the label is 0) or the right half S_{22} of S_2 (when the first digit of the label is 1). Whichever of these is correct, that half will consist of two intervals of length 3^{-2}. If x is in the leftmost of these two intervals then the second digit of the label of x is 0. Otherwise the second digit is 1. Continuing in this fashion, we may assign to x an infinite sequence of 0s and 1s.

Conversely, if $a, b, c, \ldots$ is a sequence of 0s and 1s, then we may locate a unique corresponding element y of the Cantor set. If the first digit is a zero then y is in the left half of S_1; otherwise y is in the right half of S_1. Likewise the second digit locates y within S_2, and so forth.

Thus we have a one-to-one correspondence between the Cantor set and the collection of all infinite sequences of zeroes and ones. (We are in effect thinking of the point assigned to a sequence $c_1 c_2 c_3 \ldots$ of 0s and 1s as the limit of the points assigned to

$$c_1, \; c_1 c_2, \; c_1 c_2 c_3, \; \ldots.$$

Thus we are using the fact that C is closed.) However, the set of all infinite sequences of zeroes and ones is uncountable. Thus the Cantor set is uncountable. $\quad\square$

The Cantor set is quite thin (it has zero length) but it is large in the sense that it has uncountably many elements. Also it is compact. The next result reveals a surprising, and not generally well known, property of this "thin" set:

Theorem 1.11.3. *Let C be the Cantor set and define*

$$K = \{x + y : x \in C, y \in C\}.$$

Then $K = [0, 2]$.

Proof: We sketch the proof.

Since $C \subseteq [0, 1]$ it is clear that $K \subseteq [0, 2]$. For the reverse inclusion, fix an element $t \in [0, 2]$. Our job is to find two elements c and d in C such that $c + d = t$.

We use the notation in the construction of the Cantor set. First observe that $\{x + y : x \in S_1, y \in S_1\} = [0, 2]$. Therefore there exist $x_1 \in S_1$ and $y_1 \in S_1$ such that $x_1 + y_1 = t$.

Similarly, $\{x + y : x \in S_2, y \in S_2\} = [0, 2]$. Therefore there exist $x_2 \in S_2$ and $y_2 \in S_2$ such that $x_2 + y_2 = t$.

Continuing in this fashion we may find for each j numbers x_j and y_j such that $x_j, y_j \in S_j$ and $x_j + y_j = t$. Of course $\{x_j\} \subseteq C$ and $\{y_j\} \subseteq C$ hence there are subsequences $\{x_{j_k}\}$ and $\{y_{j_k}\}$ that converge to real numbers c and d respectively. Since C is compact, we can be sure that $c \in C$ and $d \in C$. But the operation of addition respects limits, thus we may pass to the limit as $k \to \infty$ in the equation

$$x_{j_k} + y_{j_k} = t$$

to obtain

$$c + d = t.$$

Therefore $[0, 2] \subseteq \{x + y : x \in C\}$. This completes the proof.   □

Although any open set is the union of open intervals, the existence of the Cantor set shows us that there is no such structure theorem for closed sets. In fact closed intervals are atypically simple when considered as examples of closed sets.

1.12 METRIC SPACES

Certainly one of the most important examples of topological spaces is metric spaces. Metric spaces are rather special. They have more structure than most topological spaces. But they are the typical spaces for mathematical analysis so they are important. They provide a rich panoply of examples.

Definition 1.12.1. Let X be a topological space equipped with a function

$$d : X \times X \to \mathbb{R}$$

that satisfies these conditions:

 (i) $d(x, y) = d(y, x)$,

(ii) $d(x, y) \geq 0$,

(iii) $d(x, y) = 0$ if and only if $x = y$,

(iv) $d(x, y) \leq d(x, z) + d(z, y)$.

We call such a space a *metric space* and we call d the *metric*. Property **(iv)** is called the *triangle inequality*.

When a space satisfies and **(i)**, **(ii)**, and **(iv)** (but not necessarily **(iii)**) then we call it a *pseudometric space* and we call d a *pseudometric*.

EXAMPLE 1.12.2. Let $X = \mathbb{R}$ and $d(x, y) = |x - y|$. Then (X, d) is certainly a metric space. The triangle inequality is well known to any calculus student, and the other properties are elementary. The notion of distance that we have defined is the intuitive notion of distance that is familiar to any carpenter or engineer.

EXAMPLE 1.12.3. Let $X = \mathbb{R}^N$ and let

$$d(x, y) = |x - y| = \sqrt{(x_1 - y_1)^2 + \cdots + (x_N - y_N)^2}.$$

Then (X, d) is a metric space. The verification of the triangle inequality is standard, and we leave the details for the reader to verify.

EXAMPLE 1.12.4. Let $X = \mathbb{R}^N$ and let

$$d(x, y) = \begin{cases} 1 \text{ if } x \neq y \\ 0 \text{ if } x = y. \end{cases}$$

It is straightforward to check that this is a metric space.

EXAMPLE 1.12.5. Let $X = \mathbb{R}^N$ and let

$$d(x, y) = (\text{distance of } x \text{ to } \mathbf{0}) + (\text{distance of } \mathbf{0} \text{ to } y).$$

Here $\mathbf{0} = (0, 0, \ldots, 0)$ is the origin. This is known as the "New York subway metric" because the way one calculates distance from A to B in New York City is that one calculates the distance from A to Grand Central Station and then the distance from Grand Central Station to B.

Verify for yourself that d is a metric.

EXAMPLE 1.12.6. Let X be the space of all continuous functions on the interval $I = [0, 1]$. Let

$$d(f, g) = \max_{x \in I} |f(x) - g(x)|.$$

Then d is a metric on X.

First note that $d \geq 0$. Second, $d(f, g) = 0$ if and only if $f \equiv g$. Lastly, the triangle inequality for d is inherited from the triangle inequality for the real numbers.

EXAMPLE 1.12.7. Let X be the space of all continuous functions on the interval $I = [0, 1]$. Let

$$d(f, g) = |f(1/2) - g(1/2)|.$$

Then it is easy to check that this d satisfies all the axioms of a metric except **(iii)**. For if we let $f(x) = x^2/4$ and $g(x) = x^4$ then $d(f, g) = 0$ yet $f \neq g$. So this d is a pseudometric.

Definition 1.12.8. Let (X, d) be a metric space. If $x \in X$ and $r > 0$ then we let

$$B(x, r) = \{t \in X : d(x, t) < r\}.$$

We call $B(x, r)$ the *open ball* with center x and radius r. Likewise

$$\overline{B}(x, r) = \{t \in X : d(x, t) \leq r\}$$

is the *closed ball* with center x and radius r.

Remark 1.12.1. The reader may verify as an exercise that the closed ball $\overline{B}(x, r)$ is not necessarily equal to the closure of the open ball $B(x, r)$. However it will contain that closure.

Definition 1.12.9. Let (X, d) be a metric space. A set $U \subseteq X$ is said to be *open* if, for each $u \in U$, there is an $\varepsilon > 0$ such that $B(u, \varepsilon) \subseteq U$.

It is easy to verify that the open sets U specified in the last definition form a topology on X in the usual sense.

Definition 1.12.10. Let (X, d) be a metric space. A *sequence* $\{a_j\}$ in X is a function $\alpha : \mathbb{N} \to X$. We let $\alpha(1) = a_1, \alpha(2) = a_2$, etc. We say that the sequence *converges* if there is an element $\ell \in X$ such that, for every $\varepsilon > 0$, there is an $N > 0$ such that $j \geq N$ implies that $d(a_j, \ell) < \varepsilon$.

If $\{a_j\}$ is a sequence in the metric space X, then a *subsequence* is a function $A : \mathbb{N} \to \{a_j\}$ such that $\ell < k$ implies $A(\ell)$ has lower index than $A(k)$. In other words, a subsequence is an ordered list of some of the elements of the original sequence $\{a_j\}$. As an example, if the original sequence is

$$1, 2, 4, 8, 16, \ldots$$

then a subsequence is

$$1, 4, 16, \ldots .$$

Any given sequence has, in general, infinitely many (in fact uncountably many) distinct subsequences.

Theorem 1.12.2. *Let (X, d) be a metric space. A set $K \subseteq X$ is compact if and only if every sequence $\{a_j\} \subseteq K$ has a convergent subsequence (we call this last condition sequential compactness).*

Proof: Suppose that $K \subseteq X$ is compact according to our usual definition and let $\{a_j\}$ be a sequence in K. Seeking a contradiction, we assume that $\{a_j\}$ does *not* have a convergent subsequence. Then each element $k \in K$ has a neighborhood U_k that contains only finitely many elements of the sequence. But then there is a finite subcover $U_{k_1}, U_{k_2}, \ldots, U_{k_m}$. This implies that there are only finitely many elements in the sequence (as each U_{k_j} has only finitely many sequential elements). That is a contradiction.

The converse result is more work. Assume that K is sequentially compact and we want to show that it is compact (according to the original definition). First we claim that, for $\varepsilon > 0$, there is a finite subset $S = S_\varepsilon \subseteq K$ such that $K \subseteq \cup_{s \in S} B(s, \varepsilon)$. If this were not the case, then we may choose $x_1 \in K$ and then choose $x_2 \in K$ with $d(x_1, x_2) \geq \varepsilon$ and then choose

$x_3 \in K$ with both $d(x_3, x_2) \geq \varepsilon$ and $d(x_3, x_1) \geq \varepsilon$, and so forth. But then the sequence $\{x_j\}$ has no convergent subsequence, contradicting sequential compactness. We call S a *finite ε-net for K*.

Now let $\mathcal{W} = \{W_\alpha\}_{\alpha \in A}$ be an open cover of K. We claim that there is a $\delta > 0$ such that if $x \in K$ then the ball $B(x, \delta)$ lies completely inside some W_α. If this were not the case then, for each positive integer j, there is a point $x_j \in K$ such that $B(x_j, 1/j)$ does not lie in any W_α. Consider the sequence $\{x_j\}$ and let $\{x_{j_k}\}$ be a convergent subsequence. Of course the limit point ℓ of this subsequence belongs to some W_α, and there is certainly some $\delta > 0$ such that $B(\ell, \delta) \subseteq W_\alpha$. Then it is the case that, for k large enough, $B(x_{j_k}, 1/j_k) \subseteq B(\ell, \delta) \subseteq W_\alpha$. That is a contradiction. We call the number $\delta > 0$ a *Lebesgue number* (Henri L. Lebesgue (1875–1941)) for the covering $\mathcal{W}$. We say more about Lebesgue numbers in Section 1.15. We shall give an application of the concept of Lebesgue number at the end of Section 2.10.

To complete the proof, let K be sequentially compact and let $\mathcal{W} = \{W_\alpha\}_{\alpha \in A}$ be an open cover of K. Let $\delta > 0$ be a Lebesgue number for $\mathcal{W}$. Further let $\{x_1, x_2, \ldots, x_m\}$ be a finite δ-net for K. Thus for each $j = 1, \ldots, m$ there is a W_{α_j} such that $B(x_j, \delta) \subseteq W_{\alpha_j}$. Finally we see that

$$K \subseteq \bigcup_{j=1}^{m} B(x_j, \delta) \subseteq \bigcup_{j=1}^{m} W_{\alpha_j}.$$

Since the $B(x_j, \delta)$ cover K, we can thus be sure that the $\{W_{\alpha_j}\}_{j=1}^{m}$ cover K. Thus the original open cover $\mathcal{W}$ has a finite subcover. $\square$

Proposition 1.12.3. *Let (X, d) be a metric space and $f : X \rightarrow \mathbb{R}$ a function. Then f is continuous if and only if, for each $x \in X$ and each $\varepsilon > 0$, there is a $\delta > 0$ such that if $d(x, t) < \delta$ then $|f(x) - f(t)| < \varepsilon$.*

Proof: This proof is the same as that of Proposition 1.3.3. $\square$

Proposition 1.12.4. *In a metric space (X, d), the metric function d is continuous.*

Proof: In fact
$$d(x, y) \leq d(x, z) + d(z, y)$$

hence
$$d(x, y) - d(x, z) \leq d(z, y) = d(y, z).$$

By symmetry,
$$d(x, z) - d(x, y) \leq d(z, y).$$

It follows that

$$|d(x, y) - d(x, z)| \leq d(y, z).$$

In like manner,

$$|d(x, y) - d(z, y)| \leq d(x, z). \tag{1.12.4.1}$$

Thus d is actually Lipschitz continuous—this is what condition (1.12.4.1) is called. The Lipschitz condition is stronger than continuity; it is a strong form of uniform continuity. $\square$

1.13   METRIZABILITY

It is natural to ask when a given topological space can be equipped with a metric (such that the metric topology is equivalent to the original topology). Of course metric spaces are Hausdorff. So if the given topological space is not Hausdorff then it cannot be metrized. What we need is a necessary and sufficient condition for metrizability.

One of the most fundamental results about metrizability of a topological space is the next theorem of Urysohn.

Definition 1.13.1. The *Hilbert cube* is the collection of all functions from $\aleph_0$, the first infinite cardinal, to the unit interval $[0, 1]$. We can also think of the Hilbert cube as the set of all sequences $\{a_j\}$, where each $a_j \in [0, 1]$.

In practice it is convenient to think of the Hilbert cube as the set of all sequences $\{a_j\}$ where $|a_j| \leq 1/j$ for each $j = 1, 2, \ldots$. Of course this new definition of the Hilbert cube gives a space that is homeomorphic to the one specified in the enunciated definition.

Theorem 1.13.1. *Let $(X, \mathcal{U})$ be a T_1 topological space. Then the following are equivalent:*

(a) *X is regular and second countable.[4]*

(b) *X is separable[5] and metrizable.*

(c) *X can be embedded as a subspace of the Hilbert cube $I^{\aleph_0}$.*

Proof: We divide the proof into three natural parts.

(a) $\Rightarrow$ (c): Let $\mathcal{B}$ be a countable basis for X and define $\mathcal{C} = \{(U, V) : U, V \in \mathcal{B}, \overline{U} \subseteq V\}$. Of course $\mathcal{C}$ is a countable set. We know from

---

[4]We shall discuss second countable spaces in Section 2.4. A second countable space is one with a countable collection of open sets that generates the topology by way of taking unions.

[5]A separable space is one with a countable dense set—see Section 2.4.

Proposition 1.4.3 that X is in fact normal, so there is (by Urysohn's lemma) a continuous function $f_{UV} : X \rightarrow [0, 1]$ such that $f(\overline{U}) = 0$ and $f(X \setminus V) = 1$. Let $\mathcal{F} = \{f_{UV} : (U, V) \in \mathcal{C}\}$. Then $\mathcal{F}$ is countable, and the pairs in $\mathcal{C}$ will separate points from (disjoint) closed sets in X. Now Proposition 1.4.4 tells us that if I_f is a copy of $I = [0, 1]$ for each $f \in \mathcal{F}$, then the evaluation mapping $e : X \rightarrow \prod_{f \in \mathcal{F}} I_f$ defined by

$$[e(x)]_f = f(x)$$

is an embedding. Since $\mathcal{F}$ is countable, we see that $\prod_{f \in \mathcal{F}} I_f = I^{\aleph_0}$. Thus we have proved (c).

(c) $\Rightarrow$ (b): Of course $I^{\aleph_0}$ is separable and metric hence so is every subspace of $I^{\aleph_0}$.

(b) $\Rightarrow$ (a): This is obvious.   □

1.14 BAIRE'S THEOREM

One of the great triumphs of basic analysis is the Baire category theorem. This is a powerful result that is elegant in its simplicity. We begin by introducing some terminology.

Definition 1.14.1. Let $\{x_j\}$ be a sequence in the metric space (X, d). We say that $\{x_j\}$ is *Cauchy* provided that, for each $\varepsilon > 0$, there is an $N > 0$ such that whenever $j, k \geq N$ then $d(x_j, x_k) < \varepsilon$.

EXAMPLE 1.14.2. The sequence $\{1/j\}$ in the real line is certainly Cauchy. The sequence $a_j = (-1)^j$ is not Cauchy. We leave it to the reader to verify these assertions.

Definition 1.14.3. Let (X, d) be a metric space. We say that X is *complete* if, whenever $\{x_j\}$ is a Cauchy sequence in X, then there is a limit point $x_0 \in X$ so that $x_j \rightarrow x_0$.

EXAMPLE 1.14.4. The space $\mathbb{R}$, equipped with the usual Euclidean topology, is complete. Any Cauchy sequence in $\mathbb{R}$ has a limit *in the reals*. This is the fundamental, indeed the defining, property of the real number system.

The space $\mathbb{Q}$, equipped with the topology inherited from the reals, is *not* complete. For

$$3 \, , \, 3.1 \, , \, 3.14 \, , \, 3.141 \, , \, 3.1415 \, , \, 3.14159 \, , \, \ldots$$

is a Cauchy sequence which *does not* have a limit in $\mathbb{Q}$. In fact the putative limit of the sequence is the number that we usually call π. That number is real, but not rational.

Definition 1.14.5. Let (X, d) be a metric space. We say that $S \subseteq X$ is *dense* in X if the closure of S is X itself. In other words, S is dense if each metric ball $B(x, \varepsilon)$ contains an element of S.

EXAMPLE 1.14.6. The rational numbers $\mathbb{Q}$ are dense in the real numbers $\mathbb{R}$ equipped with the usual Euclidean topology.

Definition 1.14.7. Let (X, d) be a metric space. We say that $S \subseteq X$ is *nowhere dense* if the closure $\overline{S}$ of S contains no metric ball in X.

EXAMPLE 1.14.8. The integers $\mathbb{Z}$ are nowhere dense in the real numbers $\mathbb{R}$ equipped with the usual topology.

Theorem 1.14.1 (Baire). *Let (X, d) be a complete metric space. If each of the sets S_j, $j = 1, 2, \ldots$ is dense and open in X then $\cap_j S_j$ is dense (but not necessarily open) in X.*

Corollary 1.14.2. *Let (X, d) be a complete metric space. If each of the sets T_j, $j = 1, 2, \ldots$ is nowhere dense in X then $\cup_j T_j$ is nowhere dense in X.*

There is some classical terminology connected with Baire's theorem that is worth emphasizing (because it is so commonly used, and it aids in one's understanding).

Definition 1.14.9. Let (X, d) be a metric space. We say that a set $S \subseteq X$ is of *first category* if S can be written as the countable union of nowhere dense sets. All other sets are called *second category*.

Definition 1.14.10. A metric space (X, d) is called a *Baire space* if the intersection of each countable family of dense open sets is still dense.

A common and appealing restatement of Baire's theorem is that a complete metric space is of second category. Also a complete metric space is a Baire space.

Proof of Theorem 1.14.1: Let W be any nonempty open set in X. We need to show that $\cap_j S_j \cap W \neq \emptyset$.

As usual, we let $B(x, r)$ denote the open ball with center x and radius r and $\overline{B}(x, r)$ the corresponding closed ball. Since S_1 is dense, we can be sure that

$$W \cap S_1 \neq \emptyset. \tag{1.14.1.1}$$

So there exist $x_1 \in X$ and $1 > r_1 > 0$ such that

$$\overline{B}(x_1, r_1) \subseteq W \cap S_1. \tag{1.14.1.2}$$

Inductively, if $j \geq 2$ and x_{j-1}, r_{j-1} have been selected, then

$$S_j \cap B(x_{j-1}, r_{j-1}) \neq \emptyset.$$

So we can find $x_j, 1/j > r_j > 0$ such that

$$\overline{B}(x_j, r_j) \subseteq S_j \cap B(x_{j-1}, r_{j-1}). \tag{1.14.1.3}$$

Thus we have a sequence $\{x_j\} \subseteq X$. If now $\ell, m > j$, then we can be sure that x_ℓ and x_m both lie in $B(x_j, r_j)$. Thus $d(x_\ell, x_m) < 2r_j < 2/j$. Thus $\{x_j\}$ is a Cauchy sequence. Since X is complete, there is a limit point x such that $x_j \to x$.

Since x_ℓ lies in $\overline{B}(x_j, r_j)$ if $\ell > j$, we may conclude that x itself lies in $\overline{B}(x_j, r_j)$. As a result, by (1.14.1.3), x lies in S_j. Then (1.14.1.2) says that $x \in W$, which is what we wished to show. □

The next result is a classical theorem of Weierstrass. He proved that a continuous, nowhere differentiable function exists by constructing the function rather explicitly—using the theory of Fourier series. The proof that we present here is more abstract, and relies on Baire's theorem.

Proposition 1.14.3. *There is a continuous, real-valued function on the interval $I = [0, 1]$ that is nowhere differentiable on I.*

This proof is remarkable because **(i)** it is nonconstructive, **(ii)** it shows that nowhere differentiable functions are generic.

Proof of Proposition 1.14.3: Let $C(I)$ be the space of all real, continuous functions on the interval I. Equip this space with the uniform metric:

$$d(f, g) = \max_{x \in I} |f(x) - g(x)|.$$

Define $\mathcal{F}$ to be the set of functions in $C(I)$ which have a derivative at some point of I.

We claim:

(i) The space $C(I)$ is complete.

(ii) The set $\mathcal{F}$ is of first category in $C(I)$.

Proof of (i): Let $\{f_j\}$ be a Cauchy sequence in $C(I)$. We know from real analysis that there is a limit function f for this sequence that is continuous. We conclude therefore that $C(I)$ is complete.

Proof of (ii): For each $j = 1, 2, \ldots$ let us define

$$
\mathcal{F}_j = \Big\{ f \in C(I) : \text{for some } x \in [0, 1 - 1/j],
$$

$$
\text{whenever } h \in (0, 1/j], \ \left| \frac{f(x + h) - f(x)}{h} \right| \le j \Big\}.
$$

If a function $f \in C(I)$ has a derivative at some point of I, then for some j large enough, $f \in \mathcal{F}_j$. Thus $\mathcal{F} = \cup_j \mathcal{F}_j$. We can prove **(ii)** by showing that each $\mathcal{F}_j$ is closed and has no interior.

We first show that $\mathcal{F}_j$ has no interior. Let $f \in \mathcal{F}_j$ and $\varepsilon > 0$. We will find a $g \in C(I)$ such that $d(f, g) < \varepsilon$ and $g \notin \mathcal{F}_j$. The characterizing property of g is then that, for all $x \in [0, 1 - 1/j]$, there is some $h \in (0, 1/j]$ such that

$$
\left| \frac{g(x + h) - g(x)}{h} \right| > j.
$$

To construct g, first find a polynomial P such that $d(f, P) < \varepsilon/2$ (the Weierstrass approximation theorem—see Section 4.6—gives such a P). Let M be the maximum of $|P'|$ on I. Now let $Q \in C(I)$ be a piecewise linear function with the properties **(a)** $|Q(x)| \le \varepsilon/2$ for all x and **(b)** each line segment in the graph of Q has slope $M + j + 1$. See Figure 1.18.

Now define $g(x) = P(x) + Q(x)$. It is apparent from the triangle

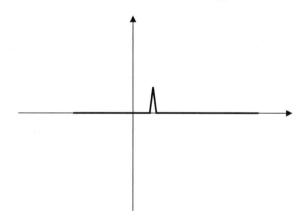

FIGURE 1.18. The function g.

inequality that $d(f, g) < \varepsilon$. Furthermore,

$$\left| \frac{g(x+h) - g(x)}{h} \right| = \left| \frac{P(x+h) + Q(x+h) - P(x) - Q(x)}{h} \right|$$

$$\geq \left| \frac{Q(x+h) - Q(x)}{h} \right| - \left| \frac{P(x+h) - P(x)}{h} \right|.$$

We know that, for $x \in [0, 1 - 1/j]$, there is an $h \in (0, 1/n]$ such that the right-hand side of the last display is $\geq (M + j + 1) - M = j + 1$. Therefore $g \notin \mathcal{F}_j$ and $\mathcal{F}_j$ has no interior.

Now we look at the closedness of $\mathcal{F}_j$. If $h_0 \in (0, 1/j]$ is a fixed element then the map $E : C(I) \times [0, 1 - 1/j]$ given by

$$E(f, x) = \left| \frac{f(x + h_0) - f(x_0)}{h_0} \right|$$

is clearly continuous. Hence $E^{-1}([0, j])$ is closed in $C(I) \times [0, 1 - 1/j]$. Let

$$D_{h_0} = \{ f \in C(I) : (f, x) \in E^{-1}([0, j]) \text{ for some } x \in [0, 1 - 1/j] \}.$$

Then D_{h_0} is closed in $C(I)$. Now let $f_\ell \in D_{h_0}$ for $\ell = 1, 2, \ldots$ with $f_\ell \to f$. Select points $x_\ell \in [0, 1 - 1/j]$ so that $(f_\ell, x_\ell) \in E^{-1}([0, j])$. Since I is compact, the sequence $\{x_\ell\}$ has a cluster point $x \in I$. Clearly $(f, x) \in E^{-1}([0, j])$, hence $f \in D_{h_0}$. We further see that

$$D_{h_0} = \left\{ f \in C(I) : \text{for some } x \in [0, 1 - 1/j], \left| \frac{f(x + h_0) - f(x)}{h_0} \right| \leq j \right\}.$$

It follows that $\mathcal{F}_j = \cap_j \{ D_{h_0} : h_0 \in (0, 1/j] \}$. So $\mathcal{F}_j$ is closed. $\square$

1.15 LEBESGUE'S LEMMA AND LEBESGUE NUMBERS

We encountered Lebesgue numbers in the proof of Theorem 1.12.2. Here we discuss the matter in more explicit detail.

Let X be a compact metric space and let $\mathcal{U} = \{U_\alpha\}_{\alpha \in A}$ be an open covering of X. We may extract a finite subcover $U_{\alpha_1}, U_{\alpha_2}, \ldots, U_{\alpha_k}$. Intuitively, we see that there is some overlap among the U_{α_j}. Our sense is that a sufficiently small ball $B(x, \varepsilon)$ must of necessity lie entirely inside one of the U_{α_j}, no matter what the location of the center x. That is the content of the next lemma of Lebesgue (Henri Lebesgue (1875–1941)).

Lemma 1.15.1. *Let X be a compact metric space and U_{α_1}, U_{α_2}, $\ldots$, U_{α_k} a finite open cover. There is a number $\varepsilon > 0$—called the Lebesgue number of the covering—so that any ball $B(x, \varepsilon)$ must lie entirely inside some U_{α_j}.*

Proof: There are many different ways to prove this result. We provide a proof by contradiction.

Suppose that the assertion is not true. Then there are points $x_\ell \in X$ and numbers $\varepsilon_\ell \searrow 0$ so that, for each ℓ, the ball $B(x_\ell, \varepsilon_\ell)$ does not lie in any U_{α_j}, $j = 1, \ldots, k$. Since X is compact, we may choose a subsequence $\{x_{\ell_m}\}$ which converges to some point x_0. Of course x_0 must lie in some U_{α_p}.

Now let $\delta = d(x_0, {}^c U_{\alpha_p}) \equiv \inf_{t \in {}^c U_{\alpha_p}} d(x_0, t)$. Certainly $\delta > 0$. If m is large enough then both

- $d(x_{\ell_m}, x_0) < \delta/2$,

- $\varepsilon_{\ell_m} < \delta/2$.

It follows from the triangle inequality that $B(x_{\ell_m}, \varepsilon_{\ell_m}) \subseteq U_{\alpha_p}$. That is a contradiction. $\square$

CHAPTER 2

ADVANCED PROPERTIES OF TOPOLOGICAL SPACES

2.1 BASIS AND SUBBASIS

We have encountered some of the ideas of this section in context in earlier parts of the book. Now we make them more formal.

Definition 2.1.1. Let $(X, \mathcal{U})$ be a topological space. We call a collection of sets $\mathcal{S} = \{S_\alpha\}_{\alpha \in A}$ a *basis* for the topology $\mathcal{U}$ if the collection of all unions of elements of $\mathcal{S}$ equals $\mathcal{U}$. Put in other words, each open set $U \in \mathcal{U}$ is the union of some elements of $\mathcal{S}$.

EXAMPLE 2.1.2. Let X be the real line with the usual topology. Then the collection of open intervals (a, b) forms a basis for this topology.

EXAMPLE 2.1.3. Let $\mathbb{R}^2$ be the plane with the usual topology. Then the collection of all open square boxes

$$S_{(x,y),\varepsilon} \equiv \{(s, t) \in \mathbb{R}^2 : |x - s| < \varepsilon, |y - t| < \varepsilon\}$$

(this is the open square box with center (x, y), sides parallel to the axes, and side length 2ε) is a basis for the topology. In fact, if U is an open set in $\mathbb{R}^2$ and if $u \in U$ then, by definition, there is an open ball $B(u, \varepsilon)$ that lies in U. But then the open square box with center u, sides parallel to the axes, and side length $\varepsilon/2$ will lie entirely in the ball, and hence entirely in U. It follows that U is the union of such boxes.

The trouble with the concept of basis is that a given collection of sets may or may not prove to be a basis for some topology. It may not be rich enough. We would like to be able to take *any* collection of sets and use it to generate a topology. That is the notion of "subbasis".

Definition 2.1.4. Let X be a space. A collection of subsets $\mathcal{T} = \{T_\alpha\}_{\alpha \in A}$ is a *subbasis* for a topology $\mathcal{U}$ on X if $\mathcal{U}$ consists of all those sets obtained from $\mathcal{S}$ through finite intersection or arbitrary union or both.

It is a simple matter to check that the process of taking finite intersection or arbitrary union, beginning with any collection of sets, will always generate a topology. That is why the concept of subbasis is a useful and flexible tool. We may rephrase the definition of subbasis as saying that $\mathcal{S}$ forms a subbasis if the family of all finite intersections of elements of $\mathcal{S}$ forms a basis.

EXAMPLE 2.1.5. Let X be the real line. The collection of all halflines $(-\infty, a)$ and (b, ∞) does not form a basis for the usual topology on X. But it does form a subbasis.

2.2 PRODUCT SPACES

Let $(X, \mathcal{U})$ and $(Y, \mathcal{V})$ be topological spaces. Then we may consider $X \times Y$ as a topological space. We use the sets $U \times V$, with $U \in \mathcal{U}$ and $V \in \mathcal{V}$, as a subbasis for the topology on the product space.

EXAMPLE 2.2.1. It would be wrong to think that the open sets in $\mathbb{R}^2$ are just the product sets

$$(a, b) \times (c, d) = \{(x, y) : a < x < b, c < y < d\}.$$

But these sets form a subbasis for the topology. In fact open sets in $\mathbb{R}^2$ can have quite arbitrary shapes—see Figure 2.1.

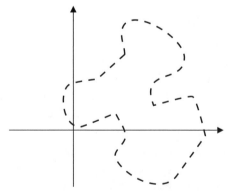

FIGURE 2.1. An open set in $\mathbb{R}^2$.

For the next fundamental theorem we will want to consider not simply finite products of spaces, but in fact infinite products—even uncountably infinite products. This will require some definitions.

Definition 2.2.2. Let $\{S_\alpha\}_{\alpha \in A}$ be sets. We define

$$\prod_{\alpha \in A} S_\alpha$$

to be the set of functions $\Phi : A \to \bigcup_\alpha S_\alpha$ such that $\Phi(\alpha) \in S_\alpha$ for each α.

To understand the last definition, if we are given $\{S_1, S_2, \ldots, S_m\}$ then the product $\prod_{j=1}^m S_j$ is defined to be the set of functions Φ from $\{1, 2, \ldots, m\}$ to $\bigcup_j S_j$ such that $\Phi(j) \in S_j$. We typically denote an element of this product by $(s_1, s_2, \ldots, s_m)$ with $s_j \in S_j$ for each j.

Definition 2.2.3. Let $\{X_\alpha\}_{\alpha \in A}$ be topological spaces. We define the projection

$$\pi_\beta : \prod_\alpha X_\alpha \to X_\beta$$

$$(x_\alpha) \mapsto x_\beta \, .$$

Then a subbasis element for the product topology is the collection of all $\pi_\alpha^{-1}(U)$ for all open subsets U in X_α, any $\alpha \in A$.

A subbasis element for the product topology as in the last definition is trivial (that is, it is the whole space) in all slots of the product except one. In the one exceptional slot it is an open set in that component space. Because of closure under finite intersection, this definition is consistent with the simpler definition that we have already considered for finite products.

Theorem 2.2.1 (Tychanoff). *The product of any number of compact spaces is compact.*

Proof: This result can be shown to equivalent to the Axiom of Choice. So using that axiom (or some equivalent thereof) in our proof will be unavoidable.

Let $\{X_\alpha\}_{\alpha \in A}$ be a family of compact topological spaces. Let

$$X = \prod_{\alpha \in A} X_\alpha \, ,$$

and equip X with the product topology. We shall use Theorem 1.5.8 and show that if $\mathscr{F} = \{F_\alpha\}_{\alpha \in A}$ is a family of closed sets in X with the finite intersection property then $\bigcap_\alpha F_\alpha$ is non-empty.

Fix an $\mathcal{F}$ as in the last paragraph. Let $\mathcal{C}$ denote the class of all families of sets in X that **(i)** contain $\mathcal{F}$ and **(ii)** have the finite intersection property. An application of Zorn's lemma, or the Hausdorff maximality principle, shows that $\mathcal{C}$ will have a maximal element.[1] Call that maximal family $\mathcal{F}'$.

The maximality of $\mathcal{F}'$ tells us that the intersection of the members of any finite subfamily of $\mathcal{F}'$ will be a set that is also an element of $\mathcal{F}'$ (or else $\mathcal{F}'$ would not be maximal).

Let $\beta \in A$, where A is the index set for the product, and let

$$\pi_\beta : X \to X_\beta$$

as usual. Define

$$\mathcal{F}'_\beta = \{\pi_\beta(F) : F \in \mathcal{F}'\}.$$

So $\mathcal{F}'_\beta$ is a family of subsets of X_β. Certainly $\mathcal{F}'_\beta$ has the finite intersection property. Since we know that X_β is compact, we may use 1.5.8 to conclude that

$$H_\beta \equiv \bigcap \left\{\overline{\pi_\beta(F)} : F \in \mathcal{F}'\right\}$$

is nonempty. Choose a point $x_\beta \in H_\beta$ for each $\beta \in A$.

Now let x denote the point in the product space X whose β^{th} coordinate is x_β. What we need to do is to prove that $x \in \overline{F}$ for every $F \in \mathcal{F}'$. That will verify the finite intersection condition, and will thus enable us to use Theorem 1.5.8 to deduce that X is compact.

So let $\widehat{U}_\beta$ be a subbasic open set in X that contains the point x. Implicit here is the fact that U_β is the projection of $\widehat{U}_\beta$ into X_β, and $U_\beta \ni x_\beta$. Moreover, the image of the projection of $\widehat{U}_\beta$ into any of the other component spaces X_α will be all of X_α. Since $x_\beta \in \overline{\pi_\beta(F)}$ for each $F \in \mathcal{F}'$, it follows that U_β meets $\pi_\beta(F)$ for each $F \in \mathcal{F}'$. Thus

$$\widehat{U}_\beta = \pi_\beta^{-1}(U_\beta)$$

meets every member F of $\mathcal{F}'$. Therefore the maximality of $\mathcal{F}'$ tells us that $\widehat{U}_\beta$ belongs to $\mathcal{F}'$, hence the intersection of any finite number of subbasic open sets containing x is a member of $\mathcal{F}'$. This implies that every neighborhood of x in X meets each member F of $\mathcal{F}'$. As a result, $x \in \overline{F}$ for each $F \in \mathcal{F}'$. We have verified the hypothesis of Theorem 1.5.8, so X is compact. $\qquad\square$

The Tychanoff theorem has many profound consequences, including the Stone-Čech compactification and the Banach-Alaoglu theorem. You will encounter these ideas in a more advanced course.

---

[1] Obviously it is here that we use the Axiom of Choice.

EXAMPLE 2.2.4. The Hilbert cube is the product $I^{\aleph_0}$ of countably many copies of the closed (compact) interval $I = [0, 1]$. (Here $\aleph_0$ is the first infinite cardinal.) In practice, in metric geometry, the Hilbert cube is often written as

$$[0, 1] \times [0, 1/2] \times [0, 1/3] \times \cdots .$$

(Certainly $[0, 1]$ is homeomorphic to $[0, 1/j]$ for each j. But the Hilbert cube as we have defined it here is easier to handle.) Of course the Hilbert cube is compact.

2.3 RELATIVE TOPOLOGY

We begin with a definition.

Definition 2.3.1. Let $(X, \mathcal{U})$ be a topological space and let $Y \subseteq X$ be a subset. We define the *relative topology* on Y to be the collection of those sets $Y \cap U$ for $U \in \mathcal{U}$. It is straightforward to check that $\{Y \cap U\}_{U \in \mathcal{U}}$ is a topology on Y.

EXAMPLE 2.3.2. Let $X = \mathbb{R}^2$ with the usual topology and consider the subspace $Y = \{(x, 0) : x \in \mathbb{R}\}$. The relative topology on Y will be the topology generated by the intervals $\{(x, 0) : a < x < b\}$.

EXAMPLE 2.3.3. Let $X = \mathbb{R}$ with the usual topology and consider the subspace $Y = [0, 1] = \{x \in \mathbb{R} : 0 \leq x \leq 1\}$. Then a basis for the relative topology on Y consists of four types of sets:

- Open intervals (a, b) with $0 < a < b < 1$,

- Half-open intervals of the form $[0, a)$ with $0 < a < 1$,

- Half-open intervals of the form $(b, 1]$ with $0 < b < 1$,

- The entire interval $[0, 1]$.

The relative topology is part of the language in this subject, and occurs throughout the rest of the book. We have seen several examples of the relative topology, defined in context, in earlier parts of the book.

2.4 FIRST COUNTABLE, SECOND COUNTABLE, AND SO FORTH

There are many different ways to measure the "size" of a set or space. One of these is based on Georg Cantor's (1845–1918) ideas of cardinality.

Definition 2.4.1. We say that a topological space $(X, \mathcal{U})$ is *separable* if it contains a countable subset $S \subseteq X$ that is dense in X. This means that every open set in X contains an element of S.

EXAMPLE 2.4.2. Let $X = \mathbb{R}$. Let $S = \mathbb{Q}$. Then S is countable, and S is dense in X. Therefore the real numbers form a separable topological space.

EXAMPLE 2.4.3. Let X be the continuous functions on the interval $[0, 1]$ with the uniform topology. Let S be the set of polynomials. Then, by the Weierstrass approximation theorem, S is dense in X (again see Section 4.6). But S is not countable, so we can make no conclusion about separability.

On the other hand, let T be those polynomials with rational coefficients. Show as an exercise that T is dense in X, and certainly T is countable. So the space X is separable.

Definition 2.4.4. Let $(X, \mathcal{U})$ be a topological space. We say that a point $x \in X$ has a *countable neighborhood base* if there is a countable collection $\{U_j^x\}$ of open sets such that every neighborhood W of x contains some U_j^x.

EXAMPLE 2.4.5. Let $X = \mathbb{R}$ with the usual topology. Then, for each $x \in \mathbb{R}$, the sets $(x - 1/j, x + 1/j)$ form a countable neighborhood base.

Definition 2.4.6. Let $(X, \mathcal{U})$ be a topological space. We say that X is *first countable* if each point $x \in X$ has a countable neighborhood base.

EXAMPLE 2.4.7. Certainly the intervals $[a, b)$ form a neighborhood base at a in the topology of the Sorgenfrey line in the reals. The intervals $[a, b')$, with b' rational, form a countable neighborhood base at a. Thus the Sorgenfrey line is first countable.

Definition 2.4.8. A topological space $(X, \mathcal{U})$ is said to be *second countable* if the topology $\mathcal{U}$ has a countable basis.

Note that first countability is a local property (at each point) while second countability is a global property.

EXAMPLE 2.4.9. Let $X = \mathbb{R}^N$ with the usual topology. If U is any open set and $u \in U$ then there is a ball $B(x, r)$ with rational center (all coordinates rational) and rational radius such that

$$u \in B(x, r) \subseteq U .$$

Thus the balls with rational center and rational radius form a basis for the topology. Therefore X is second countable.

EXAMPLE 2.4.10. Let X be the real line equipped with the topology of all sets whose complements are finite. Let $S \subseteq X$ be any infinite set. Then S is dense in X. To see this, let x be any point in X and let U be a neighborhood of x. Then the complement of U is finite, so $U \cap S \neq \emptyset$. Since S intersects every neighborhood of every point, it is dense in X. In particular $\mathbb{Z}$ is an infinite set, so it is dense. So we see that X is separable.

In spite of this, X is not second countable. To see this, suppose to the contrary that there is a countable basis for the topology of X. Fix a point x_0 of X. If U is a neighborhood of x_0 and if x is any other point of X then $U \setminus \{x\}$ is also a neighborhood of x_0. Thus there will be an open set V from the basis such that

$$x_0 \subseteq V \subseteq U \setminus \{x\}.$$

We conclude that the intersection of all basis elements that contain x_0 is simply $\{x_0\}$. If we call the basis elements U_j for $j = 1, 2, \ldots$ then we may let S_j be the complement of U_j. Of course each S_j is finite. So if U_{j_k} are the basis elements that contain x_0 then we have

$$\{x_0\} = \bigcap_k U_{j_k}$$

hence

$$\mathbb{R} \setminus \{x_0\} = \bigcup_k S_{j_k}.$$

But this says that $\mathbb{R} \setminus \{x_0\}$ is countable. That is absurd.

Proposition 2.4.1. *Any second countable space is separable.*

Proof: Let $(X, \mathcal{U})$ be the topological space. Let $\{U_j\}$ be a countable basis for the topology on X. Select a point $p_j \in U_j$ for each j. We claim that the countable set $\{p_j\}$ is dense.

For let $x \in X$ be arbitrary and U any neighborhood of x. Then some $U_j \subseteq U$, and $p_j \in U_j \subseteq U$. That does the job. $\square$

EXAMPLE 2.4.11. Let X be the real number line equipped with the discrete topology (i.e., every singleton is an open set). For any point $x \in X$, the singleton set $\{x\}$ is a neighborhood basis for the point x. So the space is first countable.

Now every set in this space is open. Therefore every set is closed. It follows that the only dense subset of X is X itself, which is certainly uncountable. Therefore X is not separable. By Proposition 2.4.1, it follows that X is not second countable.

Proposition 2.4.2. *Let X be a separable metric space. Then X is second countable.*

Proof: Let $\{p_j\}$ be a countable dense set in X. The countable base for the topology will be all metric balls with center p_j for some j and rational radius. To see this, let U be any open set and let $x \in U$. We need to produce one of the indicated balls that lies in U and contains x.

Now let $\delta > 0$ be such that $B(x, \delta) \subseteq U$. Certainly there is a p_j that lies in $B(x, \delta/3)$. Choose a rational number r that is positive and lies between $\delta/3$ and $\delta/2$. Then the ball $B(p_j, r)$ certainly contains x and, by the triangle inequality, $B(p_j, r)$ lies in U. $\qquad\qquad\square$

2.5   COMPACTIFICATIONS

As we have seen, compact spaces are often much easier to manipulate, and to prove theorems about, than arbitrary topological spaces. A useful construct is to be able to take a noncompact space and to modify it in a natural fashion to make it compact. We illustrate this idea with our first example.

EXAMPLE 2.5.1 (The Stereographic Projection). Stereographic projection puts $\widehat{\mathbb{R}^2} \equiv \mathbb{R}^2 \cup \{\infty\}$ into one-to-one correspondence with the two-dimensional sphere S in $\mathbb{R}^3$, $S = \{(x, y, z) \in \mathbb{R}^3 : x^2 + y^2 + z^2 = 1\}$, in such a way that topology is preserved in both directions of the correspondence.

In detail, begin by imagining the unit sphere bisected by the Cartesian plane with the center of the sphere $(0, 0, 0)$ coinciding with the origin in the plane—see Figure 2.2. We define the stereographic projection as follows: If $P = (x, y) \in \mathbb{R}^2$, then connect P to the north pole N of the sphere with a line segment. The point $\pi(P)$ of intersection of this segment with the sphere is called the *stereographic projection* of P. Under stereographic

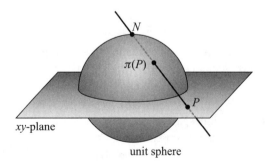

FIGURE 2.2. Stereographic projection.

projection, the point at infinity in the plane corresponds to the north pole N of the sphere. For this reason, $\mathbb{R}^2 \cup \{\infty\}$ is often thought of as being a sphere, and is then called, for historical reasons, the *Riemann sphere*.

We see that what we have done here taken the ordinary Cartesian plane and appended to it a "point at ∞." In effect, the plane (thought of as a large sheet of paper) is gathered up into a sack and the ends or edges coalesced into a point (the north pole, or point at infinity). The construction of the stereographic projection is semi-intuitive (it can be made rigorous, and one can write down an explicit formula for the mapping), but compelling. Nonetheless, it seems to be very special. If we wanted to perform a compactification on an arbitrary topological space, this would give us little hint as to how to do it.

The compactification treated in Example 2.5.1 is known as a "one-point compactification" or "Alexandroff compactification." There are in fact several different theories of compactification. We shall treat a few of them here.

First we begin with an abstract definition of what a compactification is.

Definition 2.5.2. Let $(X, \mathcal{U})$ be a non-compact topological space. A *compactification* of X is an embedding $h : X \to Y$ of X into a compact space Y so that $h(X)$ is dense in Y.

Definition 2.5.3. Let $(X, \mathcal{U})$ be a noncompact, locally compact, Hausdorff space. Let p be a point that does not lie in X. Set $X^* = X \cup \{p\}$. Let the topology $\mathcal{U}^*$ on X^* be $\mathcal{U}$ together with sets which are given by $\{p\}$ union the complement of a compact set in X. It is easy to see that $\mathcal{U}^*$ is still a topology. We call $(X^*, \mathcal{U}^*)$ a *one-point compactification* of X.

Again, Example 2.5.1 gives an example of a one-point compactification as just described. The plane is a dense subspace of the compactification (which is the sphere), because it and the sphere only differ by one point and the topology is locally Euclidean.

Proposition 2.5.1. *The compactified space X^* in Definition 2.5.3 is compact.*

Proof: Let $\mathcal{U} = \{U_\alpha\}_{\alpha \in A}$ be an open covering of X^*, with the open sets taken from the topology $\mathcal{U}^*$. Then there is at least one set that covers p. Select one of them and call it U_{α_0}. All the sets U_α that lie entirely in X cover ${}^c U_{\alpha_0}$ (which is compact), so there is a finite subcovering $U_{\alpha_1}, \ldots, U_{\alpha_k}$. Then $U_{\alpha_0}, U_{\alpha_1}, \ldots, U_{\alpha_k}$ is a finite subcovering of $\mathcal{U}$ that covers X^*. $\quad\square$

EXAMPLE 2.5.4. This is a reworking of Example 2.5.1. Let X be the Cartesian plane with the usual topology. Let p be some point that does not lie in X. Set $X^* = X \cup \{p\}$. Define a topology on X^* by taking all the usual open sets in X together with all sets of the form $^cK \cup \{p\}$, where K is a compact subset of X. Then this topologizes X^* and X^* is automatically compact by the proposition.

Another important compactification procedure is the "Stone-Čech compactification." We shall describe that now.

Let X be a Tychanoff space (Section 1.4). As at the end of Section 1.4, let C denote the set of all continuous real functions from X to the unit interval I. By the Tychanoff embedding theorem (Proposition 1.4.5), the mapping

$$f : X \to I^C$$

is an embedding and Tychanoff's compactness theorem tells us that I^C is compact. Let $\beta(X)$ denote the closure of the image $f(X)$ in I^C; certainly $\beta(X)$ is compact. Let

$$h : X \to \beta(X)$$

be the map given by f. Then $h : X \to \beta(X)$ is a compactification of X called the *Stone-Čech compactification*.

2.6   QUOTIENT TOPOLOGIES

Definition 2.6.1. Let $(X, \mathcal{U})$ be a topological space and let Y be any set. Suppose that $f : X \to Y$ is a surjective mapping. Then the collection of subsets

$$\tau_f = \{G \subseteq Y : f^{-1}(G) \text{ is open in } X\}$$

is a topology on Y called the *quotient topology* induced on Y by f. When Y is endowed with such a quotient topology, then it is called a *quotient space* of X, and the inducing map f is called a *quotient map*.

EXAMPLE 2.6.2. Let X be the standard Cartesian plane with the usual topology and let $f : X \to \mathbb{R}^3$ be the map

$$(x, y) \mapsto (\cos x, \sin x, y).$$

The image of f is a right circular cylinder in space. The quotient topology is generated by any set of the form $A \times J$, where A is any open arc of the circle and J is any open interval in $\mathbb{R}$.

Definition 2.6.3. Let $(X, \mathcal{U})$ be any topological space. A *decomposition* or *partition* of X is a pairwise disjoint family $\mathcal{Q}$ of subsets of X whose union is all of X. The function $p : X \to \mathcal{Q}$ which assigns to each element of X the unique element of $\mathcal{Q}$ to which it belongs is called the *natural projection* of X onto $\mathcal{Q}$. We endow $\mathcal{Q}$ with the quotient topology under this mapping.

Remark 2.6.1. One of the most important examples of a partition occurs when X has an equivalence relation $\sim$. The equivalence classes induced by $\sim$ form a decomposition. We let $X/\sim$ denote the collection of equivalence classes, and call it the *quotient space*.

Remark 2.6.2. In Definition 2.6.1, the sets $f^{-1}(y)$ for $y \in Y$ form a decomposition.

EXAMPLE 2.6.4. Let
$$X = \mathbb{R}^{N+1} \setminus \{0\} .$$

Say that two points p and q in X are related if there is a nonzero real number λ such that $\lambda p = q$. The resulting quotient space $X/\sim$ is called the *N-dimensional real projective space* $\mathbb{P}^N$. We see that the "points" in $\mathbb{P}^N$ are in fact the lines in $\mathbb{R}^{N+1}$ passing through the origin (with the origin of course deleted).

Let $\pi : X \to \mathbb{P}^N$ be the natural projection to the quotient. Let S^N be the unit N-dimensional sphere in $\mathbb{R}^{N+1}$. Thus S^N is a subspace of X. So we may consider
$$\pi : S^N \to \mathbb{P}^N .$$

Since every point of x can be normalized to $x/\|x\|$, which lies on the same line through the origin, we see that π (in this new form) is still surjective. In fact the map is 2-to-1: Two points a and b of S^N are mapped to the same point under π if and only if $a = -b$.

EXAMPLE 2.6.5. Let $(X, \mathcal{U})$ be a topological space, and let $E \subseteq X$ be a nonempty subset. The singletons $\{x\}$ for $x \notin E$ together with the set E form a decomposition of the space X. The resulting quotient space is called *the space obtained from X by collapsing the subset E to a point.*

As an instance, let X be the closed unit ball in $\mathbb{R}^N$:
$$X = \{(x_1, x_2, \ldots, x_N) \in \mathbb{R}^N : \sum_j |x_j|^2 \le 1\} .$$

Let S^{N-1} be the unit sphere in $\mathbb{R}^N$. Then S^{N-1} is the boundary of X. If we take the space obtained from X by collapsing its boundary to a point, we obtain a space that is homeomorphic to S^N (the unit sphere in $\mathbb{R}^{N+1}$).

2.7 Uniformities

In a metric space, the ideas of uniform convergence and uniform continuity are very natural and make good sense. On the metric space (X, d):

- The function $f : X \to \mathbb{R}$ is *uniformly continuous* if, given $\varepsilon > 0$, there is a $\delta > 0$ such that if $d(x, y) < \delta$ then $|f(x) - f(y)| < \varepsilon$.

- The sequence of functions $\{f_j\}$ with $f_j : X \to \mathbb{R}$ *converges uniformly* to a function $f : X \to \mathbb{R}$ if, given $\varepsilon > 0$, there is an $N > 0$ such that if $j \geq N$ then $|f_j(x) - f(x)| < \varepsilon$ for all $x \in X$.

The key idea here is that the distance or metric may be applied uniformly to points regardless of their location in space. A general topological space will not have such a structure, but we can impose a "uniformity" that makes the space look qualitatively like a metric space. That is the topic considered in this section.

Definition 2.7.1. Let S be any set. The *diagonal* of S, denoted $\triangle = \triangle(S)$, is the set $\{(s, s) : s \in S\}$.

Definition 2.7.2. If S and T are sets in $X \times X$ then we let

$$S \circ T = \{(s, t) : (s, u) \in T \text{ and} (u, t) \in S \text{ for some } u \in X\}.$$

This is a generalization of the notion of composition of functions.

In what follows, if $E \subseteq X \times X$, then we let $E^{-1} = \{(y, x) : (x, y) \in E\}$.

In a metric space we see that two points x and y are close to each other precisely when (x, y) is close to the diagonal. This idea motivates the following definition:

Definition 2.7.3. Let $(X, \mathcal{U})$ be a topological space. A *diagonal uniformity* on X is a collection $\mathcal{D} = \mathcal{D}(X)$ of subsets of $X \times X$ such that

(a) $D \in \mathcal{D} \Rightarrow \triangle \subseteq D$,

(b) $D_1, D_2 \in \mathcal{D} \Rightarrow D_1 \cap D_2 \in \mathcal{D}$,

(c) $D \in \mathcal{D} \Rightarrow E \circ E \subseteq D$ for some $E \in \mathcal{D}$,

(d) $D \in \mathcal{D} \Rightarrow E^{-1} \subseteq D$ for some $E \in \mathcal{D}$,

(e) $[D \in \mathcal{D}, D \subseteq E] \Rightarrow E \in \mathcal{D}$.

When X has such a structure then we call X a *uniform space*. The uniformity $\mathcal{D}$ is called *separating* if $\cap_{D \in \mathcal{D}} D = \triangle$.

A *basis* for the uniformity $\mathcal{D}$ is any subcollection $\mathcal{E}$ of $\mathcal{D}$ from which $\mathcal{D}$ can be recovered by applying condition (**e**).

EXAMPLE 2.7.4. Let X be the real numbers with the usual topology. The standard uniformity on $\mathbb{R}$ is that having as a basis the sets

$$S_\varepsilon = \{(x, y) \in \mathbb{R} \times \mathbb{R} : |x - y| < \varepsilon\}.$$

Of course, since $\mathbb{R}$ has a metric, it is very easy to write down a uniformity.

EXAMPLE 2.7.5. Let S be any set. Define $\mathcal{D}$ to be the collection of all subsets of $S \times S$ which contain $\triangle$. This is a uniformity that we call the *discrete uniformity*.

EXAMPLE 2.7.6. Let S be any set. Define $\mathcal{D}$ to be the single set $S \times S$. This uniformity on S is called the *trivial uniformity*.

We see that the uniformity in Example 2.7.5 is essentially the largest possible uniformity, while that in Example 2.7.6 is the smallest.

Definition 2.7.7. Let X be a set and $\mathcal{D}$ a uniformity on X. For $D \in \mathcal{D}$ and $x \in X$, we set

$$D[x] = \{y \in X : (x, y) \in D\}.$$

See Figure 2.3. Now if $A \subseteq X$ is any subset, we let

$$D[A] = \bigcup_{x \in A} D[x] = \{y \in X : (x, y) \in D \text{ for some } x \in A\}.$$

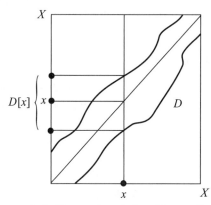

FIGURE 2.3. The topology induced by a uniformity.

Theorem 2.7.1. *Let X be a set and $\mathcal{D}$ a uniformity on X. For each $x \in X$, the collection $\mathcal{U}_x \equiv \{D[x] : D \in \mathcal{D}\}$ forms a neighborhood basis at x. Thus the uniformity induces a topology on X. This topology is Hausdorff if and only if $\mathcal{D}$ is separating.*

Proof: Certainly $x \in D[x]$ for each x. Also $D_1[x] \cap D_2[x] = (D_1 \cap D_2)[x]$, hence the intersection of neighborhoods is a neighborhood. If $D[x] \in \mathcal{U}_x$, then we know by **(c)** of the definition of uniformity that there is a set $E \in \mathcal{D}$ such that $E \circ E \subseteq D$. Thus, for any $y \in E[x]$, we see that $E[y] \subseteq D[x]$. We have confirmed all the properties of a neighborhood basis.

Suppose that $\mathcal{D}$ is separating. If x and y are distinct points of X, then there is some $D \in \mathcal{D}$ such that $(x, y) \notin D$. Then there is a symmetric $E \in \mathcal{D}$ such that $E \circ E \subseteq D$. Let $z \in E[x] \cap E[y]$. Then $(x, z) \in E$ and $(y, z) \in E$ hence $(x, y) \in E$. Therefore $(x, y) \in E \circ E \subseteq D$. This is impossible by assumption, so we may conclude that $E[x]$ and $E[y]$ are disjoint neighborhoods of x and y respectively. Thus the induced topology is Hausdorff.

For the converse to the last assertion, assume that the topology is Hausdorff. If $(x, y) \notin \Delta$ with $x \neq y$, then there are sets $S, T \in \mathcal{D}$ such that $S[x] \cap T[y] = \emptyset$. But then $S \cap T$ is an element of $\mathcal{D}$ that does not contain (x, y). □

We close this section by recording that a family $\mathcal{P}$ of pseudometrics (see Section 1.12) for a set X is called a *gage* if there is a uniformity $\mathcal{U}$ on X such that $\mathcal{P}$ is the family of all pseudometrics which are uniformly continuous on $X \times X$ relative to the product uniformity derived from $\mathcal{U}$. The idea of gage will come up later in the book.

2.8 Morse Theory

Building on earlier ideas of Arthur Cayley (1821–1895) and James Clerk Maxwell (1831–1879), Marston Morse (1892–1977) created the elegant and powerful idea of *Morse theory* (otherwise known as the *calculus of variations in the large*). The key to Morse theory is that the topology/geometry of a manifold (i.e., a surface) can be understood by examining the smooth functions (and their singularities) on that manifold.

Here we describe the fundamental ideas of basic Morse theory, but do not prove the results rigorously. The reader will come away with a good heuristic understanding of what the subject is about, and can consult a more definitive work like [MIL] for further details.

As a first example, consider a mountainous terrain in 3-dimensional Euclidean space (Figure 2.4). We examine the function, called the *Morse function*, that is *height h* of a point. Of particular interest are the singular points of the height function.

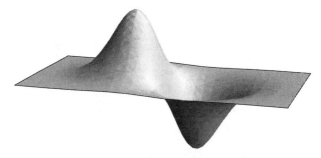

FIGURE 2.4. A mountainous terrain.

A point P on the surface will be a singular point of the height function if the gradient $\nabla h(P)$ equals 0. This means that the tangent plane to the surface will be horizontal. In Morse theory we want to consider singular points which are *nondegenerate*, meaning that the determinant of the matrix of second derivatives

$$\left(\frac{\partial^2 h}{\partial x_i \partial x_j} \right)$$

is non-zero. And we want to classify the singular point according to how many (orthogonal) tangent directions there are at the critical point in which the function h is decreasing. The number of such directions is called the *index* of the critical point.

A favorite example to help understand this classification is the torus. Look at Figure 2.5. Consider the level sets $h^{-1}(c)$ of the height function. Say that the lowest point of the torus is at height 0.

If $c < 0$ then $h^{-1}(c)$—the set of points on the torus that have height c—is the empty set, which is not very illuminating.

FIGURE 2.5. The torus.

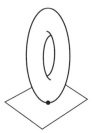

FIGURE 2.6. The first critical point on the torus.

If $c = 0$, then $h^{-1}(c)$ is a single point—see Figure 2.6. That point is a singular point, for the tangent space is horizontal; also notice that there are no directions at this point at which the height function is decreasing. In fact the height function is *increasing* in all directions, so we say that this critical point has index 0.

The basic rule of Morse theory describes how the surface changes at a critical point, and the answer depends on the index. The rule is that, if the index is γ, then the surface changes by attaching a cell of dimension γ. In our example, at the first critical point, the index is 0. So we attach a cell of dimension 0, which is a point. A point is homotopy equivalent to a disc,[2] so let us instead imagine attaching a disc. What we are doing is building the torus from scratch by examining the height function. We began with the empty set (when $c < 0$). Then we increased c until it hit the value 0. At that stage, we were at a critical point of index 0, and that signals us to paste in a 0-cell, which is a point, or (by homotopy equivalence) a disc. See Figure 2.7.

Now let c increase. For a small increasing increment, we observe that the sets $f^{-1}(c)$ are circles—see Figure 2.8. This continues until c reaches

---

[2]Two topological spaces X and Y are *homotopy equivalent* if there are mappings $f : X \to Y$ and $g : Y \to X$ with the following property: We require that there be a mapping $\Lambda_1 : X \times [0, 1] \to X$ such that

$$\Lambda_1(x, 0) = g \circ f(x)$$
$$\Lambda_1(x, 1) = \mathrm{id}_X(x)$$

and also that there be a mapping $\Lambda_2 : Y \times [0, 1] \to Y$ such that

$$\Lambda_2(y, 0) = f \circ g(y)$$
$$\Lambda_2(y, 1) = \mathrm{id}_Y(y).$$

[Here id_X and id_Y are the identity mappings on X and Y respectively.] In case X is the disc D and Y is a single point (say the origin $\mathbf{0}$) then the map $f : D \to \{\mathbf{0}\}$ is $f(\mathbf{x}) = \mathbf{0}$ and the map $g : \{\mathbf{0}\} \to D$ is $g(\mathbf{0}) = \mathbf{0}$ (where $\mathbf{0}$ is the origin of the disc in the plane). Then $\Lambda_1(\mathbf{x}, t) = t\mathbf{x}$ and $\Lambda_2(\mathbf{x}, t) = \mathbf{0}$ will do the job.

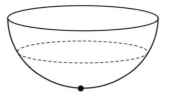

FIGURE 2.7. Building the torus by examining the critical points—the first critical point.

FIGURE 2.8. Level sets of h for c between the first two critical points.

the level of the bottom of the hole in the middle of the torus. See Figure 2.9. The point at the bottom of the hole in the middle of the torus is a critical point and $f^{-1}(c)$ is now in the shape of a figure 8. Of particular interest is that this is a saddle point. In one direction the function h is increasing and in the other direction it is decreasing, so this critical point has index 1. According to the fundamental rule of Morse theory, this means that, as c passes through the critical level at the bottom of the hole, we add a 1-cell to the surface. The result is homotopically equivalent to the surface shown in Figure 2.10.

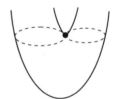

FIGURE 2.9. The second critical point.

FIGURE 2.10. The level set at the second critical point.

The next thing to notice—and this is typical in a Morse-theory analysis— is that the nature of the level sets changes when we pass through the second critical point. Before the critical point the level sets were all circles. Now the level sets are pairs of circles—see Figure 2.11.

FIGURE 2.11. Level sets after the second critical point.

The nature of the level sets will not change until we hit the next critical point. From the picture we see that the next critical point is at the *top* of the hole in the middle of the torus (see Figure 2.12). Once again, the level set at that critical point is a figure 8. The critical point is actually a saddle point, so it has index 1. That means that, as we pass through the critical point, we add a 1-cell to the surface. The result is shown in Figure 2.13. Now the level sets are once again circles.

FIGURE 2.12. The third critical point level set.

FIGURE 2.13. Structure of the surface after the third critical point.

The final critical point of the height function is at the very top of the torus. This is a critical point of index 2, as both tangential directions yield a decrease of the height function. See Figure 2.14. This means that we finish our construction by adding a 2-cell. The result is that we have built—step by step, using the analysis of critical points—the toric surface.

FIGURE 2.14. The fourth critical point, completing the construction of the torus.

One of the great theorems of nineteenth century geometry (due to August Möbius (1790–1868) and Camille Jordan (1838–1922)) is that any closed, connected surface in $\mathbb{R}^3$ (i.e., any two-manifold embedded in 3-dimensional space) is homeomorphic to a sphere with finitely many handles attached. As an example, a torus is such a surface, and it is homeomorphic to a sphere with just one handle attached. Today there are many proofs of this theorem, but one of the most charming is a proof (following the steps we have just outlined for the torus) using Morse theory.

2.9 PROPER MAPPINGS

Let X, Y be topological spaces, and let $f : X \to Y$ be a mapping. We say that f is *proper* if $f^{-1}(E)$ is compact whenever E is compact. What does this mean? First let us consider an example of a mapping that is not proper:

EXAMPLE 2.9.1. Let $X = (-1, 1)$ and let $Y = (-1, 1)$. Let $f : X \to Y$ be given by $f(x) = 1 - x^2$. Although f is continuous, it is *not* proper because $f^{-1}([0, 3/4]) = (-1, -1/2] \cup [1/2, 1)$. So we see that the inverse image of a compact set is not necessarily compact.

The next proposition helps to clarify the concept of properness:

Proposition 2.9.1. *Let X, Y be Euclidean spaces. Let $E \subseteq X$ and $F \subseteq Y$ be bounded, open sets. Suppose that $f : E \to F$ is proper. If $\{e_j\} \subseteq E$ satisfies $e_j \to \partial E$ then $f(e_j) \to \partial F$.*

Proof: If not then there are $e_j \in E$ with $e_j \to \partial E$ but $f(e_j) \not\to \partial F$. It follows that there is a compact set $K \subseteq F$ so that $f(e_j) \in K$ for all j. But then $f^{-1}(K)$ is compact, and $\{e_j\} \subseteq f^{-1}(K)$. This means that the sequence $\{e_j\}$ cannot converge to ∂E, and that is a contradiction. □

Proposition 2.9.2. *Let X, Y be topological spaces and let $f : X \to Y$ be a homeomorphism. Then f is proper.*

Proof: Now $f^{-1} : Y \to X$ is a continuous mapping. If $K \subseteq Y$ is compact then of $f^{-1}(K)$ must also be compact. So f is proper. □

Thus we see that properness is a generalization of the property of being a homeomorphism. And, at least in the familiar context of Euclidean space, a proper mapping is one that takes the boundary to the boundary. (The *Brouwer invariance of domain* theorem [HUW] generalizes this idea even further.)

Proposition 2.9.3. *Every continuous mapping f from a compact space X to a Hausdorff space Y is both proper and closed.*

Proof: The closedness is clear, for if $E \subseteq X$ is closed then it is compact. Hence $f(E)$ is compact and, since Y is Hausdorff, $f(E)$ is closed.

For the properness, let $F \subseteq Y$ be compact. Then F is certainly closed. Since f is continuous, we may be sure that $f^{-1}(F)$ is closed. But a closed subset of a compact space is compact, hence $f^{-1}(F)$ is compact. Thus f is proper. □

Proposition 2.9.4. *Let X be a topological space. Then X is compact if and only if the map of X to a single-point space Z is proper.*

Proof: Let $Z = \{z\}$. Suppose that $f : X \to Z$ is proper. Certainly Z is compact, so $f^{-1}(Z)$ is compact. But $f^{-1}(Z) = X$. That proves one direction.

Now suppose that X is compact. Let $f : X \to Z$. Let F be a compact subset of Z. Then either $F = Z$ or F is the empty set. In the first instance, $f^{-1}(F) = X$ and is compact. In the second instance $f^{-1}(F) = \emptyset$ and is compact. So f is proper.   □

2.10 PARACOMPACTNESS

Recall the concept of open cover that we used to good effect in our study of compactness (Section 1.5). Let us say that an open cover $\mathcal{U} = \{U_\alpha\}_{\alpha \in A}$ of a space X is *locally finite* if each point $x \in X$ has a neighborhood V so that V has nontrivial intersection with only finitely many of the U_α.

A *refinement* of an open cover $\mathcal{U} = \{U_\alpha\}_{\alpha \in A}$ is collection $\mathcal{V} = \{V_\beta\}_{\beta \in B}$ of open sets such that **(i)** $\mathcal{V}$ still covers X and **(ii)** each V_β is a subset of some U_α.

Definition 2.10.1. Let X be a topological space. We say that X is *paracompact* if every open cover of X admits a locally finite refinement.

Paracompactness is a generalization of compactness because every compact space is obviously paracompact. It turns out—and this is perhaps the fundamental theorem in the subject—that every metric space is paracompact. The reference [RUD] provides a particularly brief and modern proof of that result. It is also a fact (theorem of Jean Dieudonné (1906–1992)) that every paracompact space is normal.

EXAMPLE 2.10.2. One of the most famous examples in topology is the *long line*. This is uncountably many copies of the half-open unit interval pasted end-to-end. We perform the construction as follows:

Let $I = [0, 1)$. Let us consider the product $\mathcal{P} = \omega_1 \times I$, where ω_1 is the first uncountable ordinal. If $A = (s, x)$ and $B = (t, y)$ are elements of $\mathcal{P}$ then we say that $A < B$ if either $s < t$ or $s = t$ and $x < y$. This makes $\mathcal{P}$ into a totally ordered space which we call $\widehat{\mathbb{R}}$, the long line. We use sets of the form $\{X \in \widehat{\mathbb{R}} : X < A\}$ and $\{X \in \widehat{\mathbb{R}} : X > B\}$ to form a subbasis for the topology on $\widehat{\mathbb{R}}$.

We see that the long line is locally just like the real line. But it is *very* long. It is easy to see that $\widehat{\mathbb{R}}$ is not paracompact. For consider the open covering consisting of the sets $U_X = \{T \in \widehat{\mathbb{R}} : X < T\}$ for $X \in \widehat{\mathbb{R}}$. The sets $\{U_X\}$ form an open covering of $\widehat{\mathbb{R}}$. If P is any point of $\widehat{\mathbb{R}}$ and V is any neighborhood of P then V will intersect uncountably many of the U_X, and any refinement of this covering will fail to be locally finite as well.

In practice the most important property of paracompactness relates to the concept of partition of unity.

Definition 2.10.3. Let X be a topological space and let $\mathcal{U} = \{U_\alpha\}_{\alpha \in A}$ be a locally finite cover of X by open sets. We call a collection φ_α of continuous functions on X a *partition of unity subordinate to* $\mathcal{U}$ if

(i) Each φ_α satisfies $0 \le \varphi_\alpha(x) \le 1$ for all x.

(ii) For each α, the set $S_\alpha = \overline{\{x \in X : \varphi_\alpha(x) \ne 0\}}$ lies entirely in U_α.

(iii) We have the identity

$$\sum_\alpha \varphi_\alpha(x) \equiv 1 .$$

If ψ is a continuous function on a space X then we define the *support* of ψ to be the complement of the union of all open sets on which ψ vanishes. Essentially, the support of the function ψ is the set where ψ is nonzero. In the definition of partition of unity, each φ_α has support lying in U_α.

Partitions of unity are extremely useful because we can make a local construction on each U_α and then patch them together with the partition of unity. We shall give examples later.

Theorem 2.10.1. *Let X be a paracompact topological space. Let $\mathcal{U} = \{U_\alpha\}_{\alpha \in A}$ be an open cover of X. Then there is a partition of unity $\{\varphi_\alpha\}$ subordinate to $\mathcal{U}$.*

Proof: For simplicity we shall treat only the case when X is a metric space with metric d. Let $\mathcal{V} = \{V_\beta\}_{\beta \in B}$ be a refinement of $\mathcal{U}$ that is locally finite and which still covers X. For each $\beta \in B$, define

$$\psi_\beta(x) = \begin{cases} d(x, {}^c V_\beta) & \text{if } x \in V_\beta \\ 0 & \text{if } x \notin V_\beta . \end{cases}$$

Define

$$\varphi_\beta(x) = \frac{\psi_\beta(x)}{\sum_\gamma \psi_\gamma(x)} .$$

We note that each point $x \in X$ lies in some V_γ. Therefore $\psi_\gamma(x) \ne 0$ and the sum in the denominator does not vanish. Further, the covering $\mathcal{V}$ is locally finite, hence, for each fixed x, the sum in the denominator is actually finite. Also, for each x, $0 \le \varphi_\beta(x) \le 1$.

Finally, we may check that, for each $x \in X$,

$$\sum_{\beta} \varphi_{\beta}(x) = \sum_{\beta} \left[\frac{\psi_{\beta}(x)}{\sum_{\gamma} \psi_{\gamma}(x)} \right] = \frac{\sum_{\beta} \psi_{\beta}(x)}{\sum_{\gamma} \psi_{\gamma}(x)} = 1.$$

That completes the proof.   □

EXAMPLE 2.10.4. Let X be a compact metric space and let $\mathcal{U} = \{U_j\}_{j=1}^{k}$ be an open cover of X. Let $\delta > 0$ be the Lebesgue number of this cover (see Section 1.15). Then, by compactness, there is a cover of X by balls $\{B(x_j, \delta)\}_{j=1}^{k}$. We can then find a partition of unity that is subordinate to this covering by balls.

CHAPTER **3**

MOORE-SMITH CONVERGENCE AND NETS

3.1 INTRODUCTORY REMARKS

One of the nice features of the metric space setting is that all topological notions can be formulated in terms of sequences. Such is not the case in an arbitrary topological space. In that general setting we use the theory of nets and the associated idea of Moore-Smith convergence. That is the topic of the present chapter.

Whereas a sequence is modeled on the natural numbers, a net is modeled on a more general object called a directed set. The notion is similar to that for sequences, but it is more abstract. We shall find good use for nets later in the book, especially in the chapter on function spaces.

3.2 NETS

We begin by defining directed sets, and then nets.

Definition 3.2.1. Let D be a nonempty set and $\geq$ a binary relation on D. We say that $\geq$ *directs* D provided that:

(a) If m, n, p are members of D such that $m \geq n$ and $n \geq p$, then $m \geq p$.

(b) If $m \in D$, then $m \geq m$.

(c) If m, n are elements of D, then there exists $p \in D$ such that $p \geq m$ and $p \geq n$.

We note that **(a)** is a *transitivity* property and **(b)** is *reflexivity* property. Condition **(c)** might be called the *Archimedean property*. We call D or $(D, \geq)$ a *directed set*.

EXAMPLE 3.2.2. The set $\mathbb{R}$, the real numbers, with the usual ordering $\geq$, is a directed set. The nonnegative integers $\mathbb{N}$, together with the ordering $\geq$, is a directed set.

EXAMPLE 3.2.3. Let $(X, \mathcal{U})$ be an topological space. Let $x \in X$ and let $\mathcal{E}$ be the family of all neighborhoods of x. Then $\mathcal{E}$ is directed by the relation $\subseteq$.

EXAMPLE 3.2.4. Let X be any infinite set and let $\mathcal{E}$ be the collection of all finite subsets of X. Then $\mathcal{E}$ is directed by $\subseteq$.

Definition 3.2.5. A *net* is a pair $(f, \geq)$, where $f : X \to Y$ is a function and $\geq$ directs X. It is sometimes convenient to write a net as $(S_n, n \in D, \geq)$, where the directed set D is the domain of the function S (here $n \in D$ is playing the role of $x \in X$ and S takes n to S_n). When the context is understood, we shall often write the net as $(S, \geq)$.

Definition 3.2.6. We say that the net $(S_n, n \in D, \geq)$ is *in* the set A if $S_n \in A$ for all $n \in D$. We say that $(S_n, n \in D, \geq)$ is *eventually in* A if there is an element $m \in D$ such that, if $n \in D$ and $n \geq m$, then $S_n \in A$. Finally, the net is *frequently in* A if, for each $m \in D$, there is an $n \in D$ such that $n \geq m$ and $S_n \in A$.

Remark 3.2.1. If $(S_n, n \in D, \geq)$ is frequently in A then the set E of all members $n \in D$ such that $S_n \in A$ has the property that, for each $m \in D$, there is a $p \in E$ such that $p \geq m$. The reader may verify this assertion as an exercise. The subset E is called *cofinal*.

Definition 3.2.7. A net $(S, \geq)$ in a topological space $(X, \mathcal{U})$ *converges* to s relative to $\mathcal{U}$ if and only if it is eventually in every $\mathcal{U}$-neighborhood of s.

Definition 3.2.8. A net $(S, n \in D, \geq)$ on X is a *Cauchy net* with respect to a uniformity $\mathcal{D}$ on X if, for all $U \in \mathcal{D}$, there exists a $p \in D$ so that, for all $m \geq p, n \geq p$ in D, it holds that $(S_m, S_n) \in U$.

It is not difficult to build on what has been presented thus far and describe the accumulation points of a set, the closure of a set, and indeed the topology of a space in terms of convergence of nets.

Theorem 3.2.2. *Let* $(X, \mathcal{U})$ *be a topological space. Then*

(a) *A point s is an accumulation point of a subset A of X if and only if there is a net in $A \setminus \{s\}$ which converges to s.*

(b) *A point s belongs to the closure of a subset A of X if and only if there is a net in A converging to s.*

(c) *A subset A of X is closed if and only if no net in A converges to a point of $X \setminus A$.*

Proof: If s is an accumulation point of A then, for each neighborhood U of s, there is a point t_U of A that belongs to $U \setminus \{s\}$. The family $\mathcal{U}$ of all neighborhoods U of s is directed by $\subseteq$. If U and V are neighborhoods of s such that $V \subseteq U$, then $t_V \in V \subseteq U$. The net $(t_U, U \in \mathcal{U}, \subseteq)$ therefore converges to s. For the converse, if a net in $A \setminus \{s\}$ converges to s, then this net has values in every neighborhood of s and $A \setminus \{s\}$ intersects every neighborhood of s. This proves **(a)**.

To prove **(b)**, we recall that the closure of a set A consists of A together with all the accumulation points of A. For each accumulation point a of A, there is (by our discussion above) a net in A converging to a. For each point s of A, any net whose value at every element of its domain is s certainly converges to s. Therefore each point of the closure of A has a net in A converging to it. Conversely, if there is a net in A converging to s, then every neighborhood of s intersects A and s belongs to the closure of A.

Part **(c)** is now immediate. $\qquad \square$

EXAMPLE 3.2.9. A net in a topological space may converge to several different points. As an example, consider the space of integers with the topology consisting of sets whose complements are finite. Let S be the net $S_j = j$, with the domain $\mathbb{N}$ directed by the usual ordering $\geq$. Then it is easy to see that S is eventually in every open set. So S converges to every element of our topological space.

Theorem 3.2.3. *A topological space $(X, \mathcal{U})$ is Hausdorff if and only if each net in the space converges to at most one point.*

Proof: Let X be Hausdorff and let $s, t \in X$ be distinct points. Then there are disjoint neighborhoods U, V of s, t respectively. Since a net cannot eventually be in each of two disjoint neighborhoods, it is clear that no net in X converges to both points s and t.

For the converse, assume that X is not a Hausdorff space. Let s, t be distinct points of X so that every neighborhood of s intersects every neighborhood of t. If $\mathcal{U}_s$ is the collection of all neighborhoods of s and $\mathcal{U}_t$ is the collection of all neighborhoods of t then each of these collections is directed by $\subseteq$. Order the Cartesian product $\mathcal{U}_s \times \mathcal{U}_t$ by saying that $(T, U) \geq (V, W)$ provided that $T \subseteq V$ and $U \subseteq W$. Then the Cartesian product so described is directed by $\geq$. If now $(T, U) \in \mathcal{U}_s \times \mathcal{U}_t$ then the intersection $T \cap U$ is nonempty so we may select a point $p_{T,U} \in T \cap U$. If $(V, W) \geq (T, U)$ then $p_{V,W} \in V \cap W \subseteq T \cap U$. As a result, the net $(p_{T,U}, (T, U) \in \mathcal{U}_s \times \mathcal{U}_t, \geq)$ converges to both s and t. $\square$

The idea of nets, and the associated concept of convergence, is referred to in the literature as Moore-Smith convergence.

CHAPTER 4

FUNCTION SPACES

4.1 PRELIMINARY IDEAS

Many interesting examples of topological spaces arise in the context of function spaces. Function spaces are natural artifacts of analysis, and they are interesting because they are usually infinite dimensional. What does this mean?

If V is a vector space over the field $\mathbb{R}$, then V will have a basis.[1] If the basis has finitely many elements $v_1, \ldots, v_k$, then any other basis for V will also have k elements. We call k the *dimension* of V. There is also a notion of dimension for a topological space that is not necessarily a vector space. We shall not provide the details here, but refer the interested reader to [HUW] and also to our brief discussion of the idea in Section 1.10.

EXAMPLE 4.1.1. Let X be the linear space of all continuous functions on the real line. This is certainly a vector space—closed under addition and scalar multiplication. Define

$$\varphi_j(x) = \begin{cases} 0 & \text{if} \quad x \leq j \\ x - j & \text{if} \quad j < x \leq j + 1/2 \\ (j+1) - x & \text{if} \quad j + 1/2 < x \leq j + 1 \\ 0 & \text{if} \quad j + 1 < x. \end{cases}$$

Then it is plain that the functions φ_j lie in X and are linearly independent— see Figure 4.1. There are infinitely many of these functions, so that X does *not* have a finite basis. We say that X is an *infinite-dimensional* vector space.

---

[1] This statement is actually a nontrivial theorem that requires the Axiom of Choice for proof.

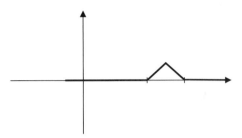

FIGURE 4.1. The functions φ_j.

A space of functions with common domain having infinitely many elements will usually be infinite dimensional. Thus there will be a great many different topologies on such a space (finite-dimensional spaces tend to have few topologies). We shall study some of them here.

4.2   THE TOPOLOGY OF POINTWISE CONVERGENCE

Let S be any set and let $\mathcal{F}(S)$ be the space of real-valued functions with domain S. We also denote this space as $\mathbb{R}^S$, and we think of this as a product space (see Section 2.2 and also [KRA]). The topology of pointwise convergence is nothing other than the product topology on $\mathbb{R}^S$. A subbasis for the topology is the collection of sets

$$\mathcal{G}_{s,U} = \{f \in \mathcal{F}(S) : f(s) \in U\}$$

for $s \in S$ a fixed point and $U \subseteq \mathbb{R}$ an open set.

In plain language, a sequence of functions f_j on S converges pointwise if, for each $s \in S$, $\lim_{j \to \infty} f_j(s)$ exists. Then we define a limit function by $f_0(s) = \lim_{j \to \infty} f_j(s)$.

Definition 4.2.1. Let $\mathcal{E}$ be a family of functions from the set S to the real numbers $\mathbb{R}$. We say that $\mathcal{E}$ is *pointwise closed* if it is closed as a subset of $\mathbb{R}^S$.

Proposition 4.2.1. *Let $\mathcal{E}$ be a family of functions from a set S to $\mathbb{R}$. Then $\mathcal{E}$ is compact with respect to the topology of pointwise convergence provided that*

(a) *The set $\mathcal{E}$ is pointwise closed in $\mathbb{R}^S$,*

(b) *for each point $s \in S$, the set $\mathcal{E}(s) \equiv \{f(s) : f \in \mathcal{E}\}$ has compact closure in $\mathbb{R}$.*

The conditions **(a)** *and* **(b)** *are also necessary for* $\mathcal{E}$ *to be compact in the topology of pointwise convergence.*

Proof: The family $\mathcal{E}$ is a subfamily of $\mathbb{R}^S$, which is also contained in $\times_{s \in S} \overline{\mathcal{E}}(s)$. If condition **(b)** holds, then the product is a compact subset of $\mathbb{R}^S$ by Tychanoff's theorem. If $\mathcal{E}$ is pointwise closed, then $\mathcal{E}$ is compact. This proves the sufficiency of **(a)** and **(b)** for compactness of $\mathcal{E}$ in the topology of pointwise convergence.

For the converse, suppose that $\mathcal{E}$ is compact in the topology of pointwise convergence. Then of course $\mathcal{E}$ is closed. The set $\mathcal{E}(s)$ is compact for each $s \in S$ and it is closed because the point evaluation map $e_s : \mathcal{E} \to \mathbb{R}$ given by $e_s(f) = f(s)$ is continuous. That proves the result. $\qquad \square$

Definition 4.2.2. Let $\mathcal{E}$ be a family of functions with common domain S. Let $A \subseteq S$. We say that the set A *distinguishes members of* $\mathcal{E}$ provided that, if f and g are distinct members of $\mathcal{E}$, then there is a point $a \in A$ such that $f(a) \neq g(a)$.

Proposition 4.2.2. *Let* $\mathcal{E}$ *be a family of functions on the set* S, *with values in* $\mathbb{R}$. *Let* $A \subseteq S$. *The family* $\mathcal{E}$ *with the topology of pointwise convergence on* A *is a Hausdorff space if and only if* A *distinguishes members of* $\mathcal{E}$.

Proof: The product space $\mathbb{R}^A$ is Hausdorff. Of course $\mathcal{E}$ with the topology of pointwise convergence on A is Hausdorff if and only if the map that takes $f \in \mathcal{E}$ to its restriction to the domain A is one-to-one. This will hold if and only if A distinguishes members of $\mathcal{E}$. $\qquad \square$

4.3 THE COMPACT-OPEN TOPOLOGY

The compact-open topology arises naturally in complex variable theory, in the study of isometry groups in differential geometry, and in other contexts as well. A motivating question for this topology is this: If $\mathcal{E}$ is a family of functions from S to $\mathbb{R}$, then under what circumstances is the mapping

$$i : \mathcal{E} \times S \to \mathbb{R}$$
$$(f, s) \mapsto f(s)$$

continuous?

Definition 4.3.1. Let $(X, \mathcal{U})$ and $(Y, \mathcal{V})$ be topological spaces. Let $\mathcal{E}$ be a family of functions from X to Y. If $K \subseteq X$ and $U \subseteq Y$, we let $\mathcal{W}(K, U)$ denote those $f \in \mathcal{E}$ such that $f(K) \subseteq U$. The family of subsets $\mathcal{W}(K, U)$

when K is compact in X and U is open in Y forms a subbasis for a topology $\mathcal{C}$ on $\mathcal{E}$ called the *compact-open topology*.

Because each singleton set is compact, it is now simple to compare the compact-open topology with the topology of pointwise convergence.

Proposition 4.3.1. *The compact-open topology $\mathcal{C}$ contains the topology $\mathcal{P}$ of pointwise convergence. The space $(\mathcal{E}, \mathcal{C})$ is a Hausdorff space provided that the range space Y is Hausdorff.*

Proof: For each $x \in X$ and each open subset $U \subseteq Y$, the set

$$\mathcal{W}(\{x\}, U) = \{f : f(x) \in U\} \tag{4.3.1.1}$$

belongs to $\mathcal{C}$ because $\{x\}$ is compact. Therefore $\mathcal{P} \subseteq \mathcal{C}$ because the family of all sets of the form (4.3.1.1) is a subbasis for the pointwise topology $\mathcal{P}$.

If Y is a Hausdorff space, then $(\mathcal{E}, \mathcal{P})$ is a Hausdorff space, because the product of Hausdorff spaces is Hausdorff. If U, V are disjoint $\mathcal{P}$-neighborhoods of distinct members f, g of $\mathcal{E}$, then they are also $\mathcal{C}$-neighborhoods. So $(\mathcal{E}, \mathcal{C})$ is Hausdorff. □

We leave it to the reader to now verify that the query posed at the beginning of this section has an affirmative answer when the space $\mathcal{E}$ of functions is equipped with the compact-open topology. See also 4.5.3 below.

4.4 UNIFORM CONVERGENCE

Here we study the concept of uniformity for a family $\mathcal{E}$ of functions from a set X to a uniform space $(Y, \mathcal{V})$. Our results do not depend on, and do not require, any topological structure on the set X. However, if X has a topology, then we shall be able to consider the question of whether the uniform limit of continuous functions is continuous.

Definition 4.4.1. Let $\mathcal{E}$ be a family of functions from a set X to a uniform space $(Y, \mathcal{V})$. For each $V \in \mathcal{V}$, let $W(V)$ be the set of all pairs of functions (f, g) such that $(f(x), g(x)) \in V$ for each $x \in X$.[2] Let $W(V)[f]$ be the set of all g such that $g(x) \in V[f(x)]$ for every $x \in X$. Then

- $W(V^{-1}) = \big(W(V)\big)^{-1}$,
- $W(U \cap V) = W(U) \cap W(V)$,
- $W(U \circ V) \supseteq W(U) \circ W(V)$.

---

[2] Remember that $\mathcal{V}$ is a uniformity, so consists of pairs. Refer to Section 2.7 for our discussion of uniformities.

for all $U, V \in \mathcal{V}$. Thus the family of sets $W(V)$ for $V \in \mathcal{V}$ is is a basis for a new uniformity $\mathcal{U}$ for $\mathcal{E}$. We call $\mathcal{U}$ the *uniformity of uniform convergence*. The topology induced by $\mathcal{U}$ is the *topology of uniform convergence*.

Theorem 4.4.1. *Let $\mathcal{E}$ be the family of all functions from a set X to a uniform space $(Y, \mathcal{V})$. Let $\mathcal{U}$ be the uniformity of uniform convergence on Y. Then*

(a) *The uniformity $\mathcal{U}$ is generated by the family of all pseudometrics of the form $d^*(f, g) = \sup\{d(f(x), g(x)) : x \in X\}$, where d is a bounded member of the gage of $(Y, \mathcal{V})$ (see Section 1.12 for the concept of pseudometric and Section 2.7 for the concept of gage).*

(b) *A net $\{f_j : j \in D\}$ converges uniformly to g if and only if it is a Cauchy net relative to $\mathcal{U}$ and $\{f_j(x), j \in D\}$ converges to $g(x)$ for each $x \in X$.*

(c) *If $(Y, \mathcal{V})$ is complete then so is the uniform space $(\mathcal{E}, \mathcal{U})$.*

Proof: For part **(a)**, if d is a bounded member of the gage of $\mathcal{V}$, then the family of all sets of the form $\{(y, z) : d(y, z) \le r\}, r > 0$, is a basis for $\mathcal{V}$. This is so because if e is a pseudometric then the pseudometric $d^* = \min\{1, e\}$ is bounded and has the same uniformity. But

$$\{(f, g) : d^*(f, g) \le r\} = \{(f, g) : d^*(f, g) \le r \text{ for each } x \in X\}$$
$$= W(\{(y, z) : d(y, z) \le r\}),$$

where W is the correspondence used above to define the uniformly continuous uniformity. In conclusion, d^* belongs to the gage of $\mathcal{U}$ and the pseudometrics of this form generate the gage. That establishes **(a)**.

The "only if" part of **(b)** is obvious. For the "if" part, suppose that a Cauchy net $\{f_j : j \in D\}$ converges pointwise to g; we must then show that it converges uniformly to g. Let V be an arbitrary, closed, symmetric member of $\mathcal{V}$. Select $m \in D$ so that, if $j \ge m$ and $k \ge m$, then $f_k(x) \in V[f_j(x)]$ for each $x \in X$. We may do this because the net is Cauchy relative to $\mathcal{U}$. Since $V[f_j(x)]$ is closed and $f_k(x)$ converges to $g(x)$, it follows that $g(x) \in V[f_j(x)]$ hence $f_j(x) \in V[g(x)]$ for each $j \ge m$ and all $x \in X$. So **(b)** is proved.

We note that **(c)** is immediate by **(b)** and the fact that the product of complete spaces is complete. □

The main properties of the uniformity for uniform convergence are capsulized in the following theorem.

Theorem 4.4.2. *Let $\mathcal{E}$ be the family of all continuous functions from a topological space X to a uniform space $(Y, \mathcal{V})$. Let $\mathcal{U}$ be the uniformity of uniform convergence. Then*

(a) *The family $\mathcal{E}$ is closed in the space of all functions from X to Y. As a result, $(\mathcal{E}, \mathcal{U})$ is complete if $(Y, \mathcal{V})$ is complete.*

(b) *The topology of uniform convergence is jointly continuous (as a function taking values in the product space $(Y, \mathcal{V})$).*

Proof: We prove part **(a)** indirectly by showing that the set of all noncontinuous functions is an open subset of the space $\mathcal{G}$ of all functions from X to Y. If f is not continuous at a point $x \in X$ then there is a member $V \in \mathcal{V}$ such that $f^{-1}[V[f(x)]]$ is not a neighborhood of x. Select a symmetric member W of $\mathcal{V}$ such that $W \circ W \circ W \subseteq V$. We note that if g is a function such that $(g(y), f(y)) \in W$ for each y, then $g \subseteq W \circ f$ and $f^{-1} \subseteq f^{-1} \circ W^{-1} = f^{-1} \circ W$ and therefore $g^{-1} \circ W \circ g \subseteq f^{-1} \circ W \circ W \circ W \circ f \subseteq f^{-1} \circ V \circ f$. As a result, $g^{-1}[W[g(x)]]$ is a subset of $f^{-1}[V[f(x)]]$ and is thus not a neighborhood of x. Thus **(a)** is proved.

For **(b)**, we need to demonstrate the continuity of the map of $\mathcal{E} \times X$ into Y at a point (f, x). We note that, for $V \in \mathcal{V}$, if $y \in f^{-1}[V[f(x)]]$ and $g(z) \in V[f(z)]$ for all z, then $g(y) \in V[f(y)] \subseteq V \circ V[f(x)]$. That completes the argument. $\square$

It is a useful device to consider uniform convergence on each member of a family $\mathcal{A}$ of subsets of the domain space X. For example, in complex variable theory we commonly consider the topology of uniform convergence on compact sets. In detail, if $\mathcal{E}$ is a family of functions from a set X to a uniform space $(Y, \mathcal{V})$ and if $\mathcal{A}$ is a family of subsets of X, then the uniformity of *uniform convergence on members of $\mathcal{A}$*, abbreviated $\mathcal{U}|\mathcal{A}$, has for a subbasis the family of all sets of the form

$$\{(f, g) : (f(x), g(x)) \in V \text{ for all } x \in A\}$$

where $V \in \mathcal{V}$ and $A \in \mathcal{A}$.

4.5   EQUICONTINUITY AND THE ASCOLI-ARZELA THEOREM

Equicontinuity is a notion of uniform continuity over a family of functions. It is useful when we want to prove compactness theorems for families of functions.

Definition 4.5.1. Let $\mathcal{E}$ be a family of mappings from a topological space X into a uniform space $(Y, \mathcal{V})$. The family $\mathcal{E}$ is said to be *equicontinuous at a point* x if, for each $V \in \mathcal{V}$, there is a neighborhood U of x such that $f[U] \subseteq V[f(x)]$ for every $f \in \mathcal{E}$.

Remark 4.5.1. In case (X, d), (Y, e) are metric spaces then there is a particularly elegant and compelling formulation of equicontinuity. We say that a family of functions $f_\alpha : X \to Y$ is equicontinuous at x if, for any $\varepsilon > 0$, there is a $\delta > 0$ such that if $d(x, t) < \delta$ then $e(f_\alpha(x), f_\alpha(t)) < \varepsilon$ for all α. The point is that the choice of $\delta > 0$ is independent of x and t.

Proposition 4.5.2. *If the family $\mathcal{E}$ of functions from X to Y is equicontinuous at x, then the closure of $\mathcal{E}$ relative to the topology $\mathcal{P}$ of pointwise convergence is also equicontinuous at x.*

Proof: Let $x \in X$ and U a neighborhood of x. Let V be a member of the uniformity of Y that is a closed set. Then the class of all functions f that satisfy $f[U] \subseteq V[f(x)]$ is closed relative to the topology $\mathcal{P}$ of pointwise convergence because the set is just the same as

$$\bigcap \{\{f : (f(y), f(x)) \in V\} : y \in U\}.$$

It follows that the pointwise closure of $\mathcal{E}$ is equicontinuous. $\square$

Definition 4.5.2. A family $\mathcal{E}$ of functions is said to be *equicontinuous* if it is equicontinuous at each point.

It follows that the closure of an equicontinuous family in the topology of pointwise convergence is also equicontinuous.

Proposition 4.5.3. *Let $\mathcal{E}$ be an equicontinuous family of functions. Then the topology of pointwise convergence on $\mathcal{E}$ is jointly continuous and therefore coincides with the topology of uniform convergence on compact sets.*

Proof: We want to show that the map of $\mathcal{E} \times X \to Y$ given by $(f, x) \mapsto f(x)$ is continuous at a point (f, x). Let V be a member of the uniformity of Y and let U be a neighborhood of x such that $g[U] \subseteq V[g(x)]$ for all $g \in \mathcal{E}$. If g is a member of the $\mathcal{P}$-neighborhood $\{h : h(x) \in V[f(x)]\}$ of f and $y \in U$, then $g(y) \in V[g(x)]$ and $g(x) \in V[f(x)]$. As a result, $g(y) \in V \circ V[f(x)]$ and joint continuity follows. One may verify that each jointly continuous topology is larger than the compact-open topology, and the compact-open topology coincides with that of uniform convergence on compact sets. $\square$

One interpretation of this last result is that an equicontinuous family of functions is compact relative to the topology of uniform convergence on compact sets if it is compact relative to pointwise convergence. Also the Tychanoff theorem gives suffecient criteria for compactness in the pointwise topology. Thus we see that equicontinuity plus some other conditions gives compactness for a family of functions. The next result is a sort of converse.

Theorem 4.5.4. *Let $\mathcal{E}$ be a family of functions from a topological space X to a uniform space $(Y, \mathcal{V})$. If $\mathcal{E}$ is compact relative to a jointly continuous topology, then it is equicontinuous.*

Proof: Let $x \in X$ be a fixed piont and V a symmetric member of $\mathcal{V}$. If we can show that there is a neighborhood U of x such that $g[U] \subseteq V \circ V[g(x)]$ for each g in $\mathcal{E}$, then the result will follow.

Because the topology on $\mathcal{E}$ is jointly continuous, there is for each $f \in \mathcal{E}$ a neighborhood $\mathcal{G}$ of f and a neighborhood W of x such that $\mathcal{G} \times W$ maps into $V[f(x)]$. If $g \in \mathcal{G}$ and $w \in W$, then $g(x)$ and $g(w)$ both belong to $V[f(x)]$ and hence $g(w) \in V \circ V[g(x)]$. That is to say, $g[W] \subseteq V \circ V[g(x)]$ for each $g \in \mathcal{G}$. Since $\mathcal{E}$ is compact, there is a finite family of open sets $\mathcal{G}_1, \mathcal{G}_2, \ldots, \mathcal{G}_k$ covering $\mathcal{E}$ and corresponding neighborhoods $W_1, \ldots, W_k$ of x such that $g[W_j] \subseteq V \circ V[g(x)]$ for each $g \in \mathcal{G}_j$. If we let U be the intersection of the neighborhoods W_j of x, then it is clear that $g[U] \subseteq V \circ V[g(x)]$ for every $g \in \mathcal{G}$. $\qquad \square$

Now we have the celebrated Ascoli-Arzela theorem.

Theorem 4.5.5. *Let $\mathcal{C}$ be the family of all continuous functions from a regular, locally compact topological space X to a Hausdorff uniform space $(Y, \mathcal{V})$. Assume that $\mathcal{C}$ has the topology of uniform convergence on compact sets. Then a subfamily $\mathcal{E}$ of $\mathcal{C}$ is compact if and only if*

(a) $\mathcal{E}$ *is closed in $\mathcal{C}$,*

(b) $\mathcal{E}[x]$ *has compact closure (i.e., its closure is compact) for each $x \in X$,*

(c) *the family $\mathcal{E}$ is equicontinuous.*

Proof: This is immediate from what went before—especially 4.5.3, 4.5.4.

$\qquad \square$

For clarity and precision, we now enunciate a version of Ascoli-Arzela that can be found in many textbooks (see [KRA1], [RUD]). We leave its verification to the reader.

Theorem 4.5.6. *Let X be a compact metric space. Let $\mathcal{F} = \{f_\alpha\}_{\alpha \in A}$ be a family of real-valued functions on X that satisfies*

(a) *The family $\mathcal{F}$ is equicontinuous,*

(b) *The family $\mathcal{F}$ is uniformly bounded in the sense that there is an $M > 0$ such that $|f_\alpha(x)| \leq M$ for all $f_\alpha \in \mathcal{F}$ and all $x \in X$.*

Then there is a subsequence $\{f_{\alpha_j}\}$ that converges uniformly on X to some continuous function f_0 on X.

4.6 THE WEIERSTRASS APPROXIMATION THEOREM

One of the startling results of nineteenth-century analysis was the celebrated approximation theorem of Weierstrass. A bit of context is in order so that we may appreciate the significance and meaning of the theorem.

Perhaps the simplest and easiest of all functions to understand is the polynomial, a function of the form

$$p(x) = a_0 + a_1 x + a_2 x^2 + \cdots + a_k x^k .$$

The nice thing about a polynomial is that its values are easy to calculate using simple arithmetic operations. Most functions are not like that. Even the familiar sine and cosine functions can only be calculated (by hand) at certain special values, and the same is true for the logarithm and exponential functions.

What Weierstrass tells us is that any continuous functions f on the interval $[0, 1]$ can be approximated by a polynomial. Even more striking is that this approximation is uniform: Given an $\varepsilon > 0$ there is a polynomial p such that

$$|f(x) - p(x)| < \varepsilon$$

for all $x \in [0, 1]$.

Weierstrass's ideas grew out of his studies of trigonometric and Fourier series. His proof fits naturally into that context. The argument that we present here is a modern and streamlined treatment that is more self-contained.

The name Weierstrass has occurred frequently in this chapter. Karl Weierstrass (1815–1897) revolutionized analysis with his examples and theorems. This section is devoted to one of his most striking results.

It is natural to wonder whether the standard functions of calculus— $\sin x, \cos x$, and e^x, for instance—are actually polynomials of some very high degree. Since polynomials are so much easier to understand than transcendental functions, an affirmative answer would simplify mathematics. A moment's thought shows that this wish is impossible: a polynomial of degree k has at most k real roots, and because the sine and cosine functions have infinitely many real roots they cannot be polynomials. A polynomial of degree k has the property that if it is differentiated $k + 1$ times then the resulting derivative is zero. Since this is not the case for e^x, we conclude that e^x cannot be a polynomial.

In calculus we learned of a formal procedure, called Taylor series, for associating polynomials with a given function f. In some instances these polynomials form a sequence that converges to the original function. The method of the Taylor expansion has no hope of working unless f is infinitely differentiable. Even then, it turns out that the Taylor series rarely converges to the original function. Even when the Taylor series converges, there is no guarantee that it will converge to the original f. See [KRP1] for a detailed consideration of these matters.

Nevertheless, Taylor's theorem with remainder might cause us to speculate that any reasonable function can be approximated in some fashion by polynomials. In fact the theorem of Weierstrass gives a spectacular affirmation of this intuition:

Theorem 4.6.1 (The Weierstrass Approximation Theorem). *Let f be a continuous function on an interval $[a, b]$. Then there is a sequence of polynomials $p_j(x)$ with the property that the sequence $\{p_j\}$ converges uniformly on $[a, b]$ to f.*

Before proving this theorem, let us consider some of its consequences. A restatement of the theorem would be that, given a continuous function f on $[a, b]$ and an $\varepsilon > 0$, there is a polynomial p such that

$$|f(x) - p(x)| < \varepsilon$$

for every $x \in [a, b]$. If one were programming a computer to calculate values of a fairly wild function f, the theorem guarantees that, up to a given degree of accuracy, one could use a polynomial instead; this would in fact be much easier for the computer to handle. Advanced techniques can even tell what degree of polynomial is needed to achieve a given degree of accuracy. The proof that we shall present also suggests how this might be done.

Let f be the Weierstrass nowhere differentiable function. The theorem guarantees that, on any compact interval, f is the uniform limit of polynomials. Thus even the uniform limit of infinitely differentiable functions need not be differentiable—not even at one point.

We shall break up the proof of the Weierstrass Approximation Theorem into a sequence of lemmas.

Lemma 4.6.2. *Let ψ_j be a sequence of continuous functions on the interval $[-1, 1]$ with the following properties:*

(i) $\psi_j(x) \geq 0$ *for all x,*

(ii) $\int_{-1}^{1} \psi_j(x) \, dx = 1$ *for each j,*

(iii) *For any $\delta > 0$ we have*

$$\lim_{j \to \infty} \int_{\delta \leq |x| \leq 1} \psi_j(x) \, dx = 0.$$

If f is a continuous function on the real line which is identically zero off the interval $[0, 1]$ then the functions $f_j(x) = \int_{-1}^{1} \psi_j(t) f(x - t) \, dt$ converge uniformly on the interval $[0, 1]$ to $f(x)$.

Proof: By multiplying f by a constant we may assume that $\sup |f| = 1$. Let $\varepsilon > 0$. Since f is uniformly continuous on the interval $[0, 1]$ we may choose a $\delta > 0$ such that if $|x - t| < \delta$ then $|f(x) - f(t)| < \varepsilon/2$. By property (iii), we may choose an N so large that $j > N$ implies that $|\int_{\delta \leq |t| \leq 1} \psi_j(t) \, dt| < \varepsilon/4$. Then, for any $x \in [0, 1]$, we have

$$\left| f_j(x) - f(x) \right| = \left| \int_{-1}^{1} \psi_j(t) f(x - t) \, dt - f(x) \right|$$

$$= \left| \int_{-1}^{1} \psi_j(t) f(x - t) \, dt - \int_{-1}^{1} \psi_j(t) f(x) \, dt \right|.$$

In the last line, we have used (ii) to multiply the term $f(x)$ by 1 in a clever way. We may combine the two integrals to find that

$$\left| f_j(x) - f(x) \right| = \left| \int_{-1}^{1} (f(x - t) - f(x)) \psi_j(t) \, dt \right|$$

$$\leq \int_{-\delta}^{\delta} |f(x - t) - f(x)| \psi_j(t) \, dt$$

$$+ \int_{\delta \leq |t| \leq 1} |f(x - t) - f(x)| \psi_j(t) \, dt$$

$$= A + B.$$

To estimate A, we recall that, for $|t| < \delta$, we have $|f(x - t) - f(x)| < \varepsilon/2$, so

$$A \leq \int_{-\delta}^{\delta} \frac{\varepsilon}{2} \psi_j(t) \, dt \leq \frac{\varepsilon}{2} \cdot \int_{-1}^{1} \psi_j(t) \, dt = \frac{\varepsilon}{2}.$$

For B we write

$$B \leq \int_{\delta \leq |t| \leq 1} 2 \cdot \sup |f| \cdot \psi_j(t) \, dt$$

$$\leq 2 \cdot \int_{\delta \leq |t| \leq 1} \psi_j(t) \, dt$$

$$< 2 \cdot \frac{\varepsilon}{4} = \frac{\varepsilon}{2},$$

where in the penultimate line we have used the choice of j. Adding together the estimates for A and B, and noting that they are independent of the choice of x, yields the result. $\square$

Lemma 4.6.3. *Define $\psi_j(t) = k_j \cdot (1 - t^2)^j$, where the positive constants k_j are chosen so that $\int_{-1}^{1} \psi_j(t) \, dt = 1$. Then the functions ψ_j satisfy the properties* **(i)**–**(iii)** *of the last lemma.*

Proof: Property **(ii)** is true by design. Property **(i)** is obvious. In order to verify property **(iii)**, we need to estimate the size of k_j.

We have

$$\int_{-1}^{1} (1 - t^2)^j \, dt = 2 \cdot \int_{0}^{1} (1 - t^2)^j \, dt$$

$$\geq 2 \cdot \int_{0}^{1/\sqrt{j}} (1 - t^2)^j \, dt$$

$$\geq 2 \cdot \int_{0}^{1/\sqrt{j}} (1 - jt^2) \, dt \, ,$$

where we have used the binomial theorem. The last integral is easily evaluated and equals $4/(3\sqrt{j})$. We conclude that

$$\int_{-1}^{1} (1 - t^2)^j \, dt > \frac{1}{\sqrt{j}} \, .$$

As a result, $k_j < \sqrt{j}$.

To verify property **(iii)** of the lemma, we notice that, for $\delta > 0$ fixed and $\delta \leq |t| \leq 1$, we have

$$|\psi_j(t)| \leq k_j \cdot (1 - \delta^2)^j \leq \sqrt{j} \cdot (1 - \delta^2)^j$$

and the last expression tends to 0 as $j \to \infty$. Thus $\psi_j \to 0$ uniformly on $\{t : \delta \leq |t| \leq 1\}$. It follows that the ψ_j satisfy property **(iii)** of the lemma.
                                                                      $\square$

Proof of the Weierstrass Approximation Theorem: We may assume without loss of generality (just by changing coordinates) that f is a continuous function on the interval $[0, 1]$. After adding a linear function, which is a polynomial, to f, we may assume that $f(0) = f(1) = 0$. Thus f may be continued to be a continuous function which is identically zero on the entire real line.

Let ψ_j be as in Lemma 4.6.3 and form f_j as in Lemma 4.6.2. Then we know that $\{f_j\}$ converges uniformly on $[0, 1]$ to f. Finally,

$$
\begin{aligned}
f_j(x) &= \int_{-1}^{1} \psi_j(t) f(x - t) \, dt \\
&= \int_{0}^{1} \psi_j(x - t) f(t) \, dt \\
&= k_j \int_{0}^{1} (1 + (x - t)^2)^j \, f(t) \, dt \, .
\end{aligned}
$$

Multiplying out the expression $(1 + (x - t)^2)^j$ in the integrand then shows that f_j is a polynomial of degree at most $2j$ in x. Thus we have constructed a sequence of polynomials f_j that converges uniformly to f on the interval $[0, 1]$. $\qquad\square$

TABLE OF NOTATION

Notation	Section	Definition
$\setminus$	1.2	set-theoretic difference
$\cup$	1.2	union of sets
$\cap$	1.2	intersection of sets
$(X, \mathcal{U})$	1.2	topological space
$\mathcal{U}$	1.2	topology
X	1.1	space
$(a, b), [a, b], [a, b), (a, b]$	1.2	intervals
U	1.2	open set
$\mathbb{R}$	1.2	real numbers
(a, b)	1.1	interval
$\mathbb{R}^N$	1.2	Euclidean space
$B(x, \varepsilon)$	1.2	open ball
$\overline{B}(x, \varepsilon)$	1.2	closed ball
$\mathbb{Z}$	1.2	integers
E	1.2	closed set
$^c S$	1.2	complement of S
$\overset{\circ}{S}$	1.2	interior of S
∂S	1.2	boundary of S
$\overline{S}$	1.2	closure of S
$f^{-1}(S)$	1.3	inverse image of S under f
$\mathbb{Q}$	1.3	rational numbers
$f : X \to Y$	1.3	mapping

Notation	Section	Definition
T_0 space	1.4	space satisfying the zeroeth separation axiom
T_1 space	1.4	space satisfying the first separation axiom
T_2 space	1.4	space satisfying the second separation axiom
T_3 space	1.4	space satisfying the third separation axiom
T_4 space	1.4	space satisfying the fourth separation axiom
$\mathcal{P}$	1.4	Moore plane
$\mathcal{W} = \{W_\alpha\}_{\alpha \in A}$	1.4	open covering
$\prod_\alpha X_\alpha$	1.4	product of spaces
S^T	1.4	set of functions from T to S
K	1.5	compact set
$f(E)$	1.7	the image of E under f
S	1.7	topologists's sine curve
$\gamma : [0,1] \to X$	1.8	path
(p, U, V)	1.9	cutting
$\mathbf{C}$	1.10	Cantor set
S_j	1.11	sets used to construct Cantor set
$d(x, y)$	1.12	a metric
$\mathbf{0}$	1.12	the origin
$B(x, r)$	1.12	open ball in a metric space
$\overline{B}(x, r)$	1.12	closed ball in a metric space
$\{a_j\}$	1.12	sequence
$\{a_{j_k}\}$	1.12	subsequence
$\aleph_0$	1.13	first infinite cardinal
$\mathcal{S}$	2.1	basis
$\mathcal{T}$	2.1	subbasis
$X \times Y$	2.2	product space
$\prod_{\alpha \in A} S_\alpha$	2.2	product
π_β	2.2	projection on j^{th} factor
$\{U_j^x\}$	2.4	neighborhood basis of x
$\widehat{\mathbb{R}^2}$	2.5	Riemann sphere
∞	2.5	point at infinity
X^*	2.5	one-point compactification of X
$\mathcal{U}^*$	2.5	topology of the one-point compactification

Notation	Section	Definition
$\beta(X)$	2.5	Stone-Čech compactification
τ_f	2.6	quotient topology
p	2.6	natural projection of X onto Q
$X/\sim$	2.6	quotient space
$\triangle = \triangle(S)$	2.7	diagonal of S
$S \circ T$	2.7	composition of two product sets
E^{-1}	2.7	inverse of a product set
$\mathcal{U}$	2.7	a uniformity
$\mathcal{D}$	2.7	diagonal uniformity
$D[x]$	2.7	diagonal slice at x
$D[A]$	2.7	diagonal slice at A
$\mathcal{P}$	2.7	gage
h	2.8	height function
ω_1	2.10	first uncountable ordinal
$\widehat{\mathbb{R}}$	2.10	long line
$\{\varphi_\alpha\}$	2.10	partition of unity
$(f, \geq)$	3.2	net
$(S_n, n \in D, \geq)$	3.2	net
V	4.1	vector space
$\mathcal{F}(S)$	4.2	real-valued functions with domain S
$\mathcal{G}_{s,U}$	4.2	subbasis for topology on $\mathcal{F}(S)$
$\mathcal{E}$	4.2	family of functions from S to $\mathbb{R}$
$\mathcal{W}(K, U)$	4.3	subbasis for compact-open topology
d^*	4.4	a pseudometric
$V[a]$	4.4	those b such that $(a, b) \in V$
$p(x)$	4.6	a polynomial
$\{\psi_j\}$	4.6	an approximation to the identity

GLOSSARY

accumulation point of a set A A point s with the property that some sequence or net in A converges to s. Equivalently, every neighborhood of s contains points of A.

affine mapping   A map f on Euclidean space that satisfies

$$f((1-t)P + tQ) = (1-t)f(P) + tf(Q).$$

Ascoli-Arzela theorem   The result that asserts that an equicontinuous, equibounded family of functions on a compact metric space is itself compact.

Baire category theorem   The result that a complete metric space cannot be written as the countable union of nowhere dense sets.

Baire space   A space with the property that the countable intersection of dense, open sets is still dense.

ball   Given a metric space (X, d), the ball $B(P, r)$ with center P and radius r is the set $B(P,r) = \{x \in X : d(x, P) < r\}$.

Banach-Alaoglu theorem   The result that the unit ball in the dual of a Banach space is weak-$*$ compact.

basis for a topology   A family of open sets that generates the entire topology by way of union.

basis of a vector space   A linearly independent set that generates the space via linear combinations.

boundary   A chain that is in the image of the boundary operator.

boundary point of a set   A point such that any neighborhood contains elements of the set and elements of the complement.

calculus of variations in the large   See *Morse theory*.

Cantor set   A compact set in the real line, based on a triadic decomposition process, with remarkable topological characteristics.

Cauchy net   A net that has properties analogous to a Cauchy sequence.

Cauchy sequence   A sequence $\{a_j\}$ on a metric space (X, d) so that, for each $\varepsilon > 0$, there is an $N > 0$ so that when $j, k \geq N$ then $d(a_j, a_k) < \varepsilon$.

closed set   A set E in a topological space X whose complement is open.

closure of a set   The intersection of all closed sets that contain the given set.

cofinal   For a net that is frequently in the set A, the set of indices of elements that lie in A.

collapsing a set to a point   A device, using the quotient construction, for identifying a set with a point.

compactification   A compact space that is built from a noncompact space.

compact-open topology   Let $\mathcal{E}$ be a family of functions from X to Y. For each compact $K \subseteq X$ and each open $U \subseteq Y$, let $\mathcal{W}(K, U)$ be those functions $f \in \mathcal{E}$ that map K into U. The compact-open topology is that generated by the collection of $\mathcal{W}(K, U)$ as a subbasis.

compact set   A set with the property that every open covering has a finite subcovering.

completely regular space   A topological space in which a disjoint point and closed set can be separated by a continuous function.

connected component   The equivalence classes in a space induced by the connectivity property.

connected set   A set that cannot be decomposed into two disjoint open sets.

continuous mapping   A mapping with the property that the inverse image of any open set is open.

continuum   A compact, connected Hausdorff space.

contractible space   A space X is contractible if the identity map on X is homotopic to the trivial point-map.

convergent net   A net converges to a point s if it is eventually in every neighborhood of s.

countable neighborhood base For each point $x \in X$, this is a countable collection of open sets that generates the topology in a neighborhood of x.

covering of a set A collection of sets whose union contains a given set.

decomposition of a space A partition of the space.

diagonal of a set S The set $\{(s, s) : s \in S\}$.

diagonal uniformity The basis for the topology of a uniform space.

dimension A topological invariant defined inductively in terms of the characteristics of the boundary of a set.

dimension of a vector space The number of elements in any basis for the vector space.

directed set A set equipped with a special type of order relation. The idea is used to study nets.

disconnected set A set that can be decomposed into two disjoint open sets.

discrete uniformity The largest uniformity on a space.

distinguishes members of a family of functions A set A distinguishes members of a family $\mathcal{F}$ of functions if, whenever $f, g \in \mathcal{F}$ are distinct, then there is an $a \in A$ such that $f(a) \neq g(a)$.

embedding A continuous, one-to-one mapping with a continuous inverse (on the image).

equibounded family of functions A family of functions that satisfies a uniform global bound.

equicontinuous family of functions A family of functions that is uniformly continuous over all elements of the family.

equivalent coverings Two coverings of the same base space with a morphism that commutes with the covering maps.

eventually in A A net that, after a certain index, lies in A.

finite-dimensional vector space A vector space with a finite basis.

finite intersection property A family of sets $\mathcal{F}$ has the finite intersection property if each finite subcollection of $\mathcal{F}$ has nonempty intersection.

finite subcovering A subcovering with finitely many elements.

first countable A topological space in which each point has a countable neighborhood base.

fixed point If $f : X \to X$ is a mapping then x is a fixed point of f if $f(x) = x$.

frequently in A A net that, arbitrarily far out in the directed indexing set, has elements that lie in A.

function A function $f : X \to Y$ is a set $\mathcal{F}$ of ordered pairs (x, y) with $x \in X$ and $y \in Y$ such that **(i)** if $x \in X$ then there exists a $y \in Y$ such that $(x, y) \in \mathcal{F}$ and **(ii)** if $(x, y_1) \in \mathcal{F}$ and $(x, y_2) \in \mathcal{F}$ then $y_1 = y_2$.

gage A family of pseudometrics that is uniformly continuous with respect to a product uniformity.

Hausdorff space A space that satisfies the T_2 separation axiom.

Heine-Borel theorem The result that, in Euclidean space, a compact set is a set that is closed and bounded (and conversely).

homeomorphism A one-to-one, onto, bicontinuous mapping of topological spaces.

index of a singular point The number of orthogonal tangent directions in which the Morse function is decreasing.

infinite-dimensional vector space A vector space with no finite basis.

interior point of a set A point with a neighborhood that lies entirely in the set.

Lebesgue lemma The result that a finite open cover of a compact metric space has a ball of fixed radius that will fit into each of the covering sets.

Lebesgue number The radius of the ball in the Lebesgue lemma.

Lindelöf space A topological space with the property that every open cover has a countable subcover.

linearly independent set of points A set of points in Euclidean space that has the maximal number of degrees of freedom in its coordinates.

linearly independent set of vectors A collection of vectors with no non-trivial linear combination that equals zero.

locally compact space A space with the property that each point has a neighborhood base consisting of compact sets.

long line   A topological space that consists of uncountably many half-open unit intervals pasted together.

mapping   A function that does not take values in a scalar field. In topology we often assume automatically that a mapping is continuous.

metric   The distance function on a metric space.

metric space   A space equipped with a notion of distance.

metrizable   A topological space whose topology is equivalent to a metric topology.

Moore plane   A special topology on the closed upper halfplane that gives an example of a T_3 space that is not T_4.

Moore-Smith convergence   The theory of convergence using nets.

Morse function   The function that is used as a tool in Morse theory to create level sets for understanding the geometry of a manifold.

Morse theory   Also called the calculus of variations in the large, this is a method of studying the geometry of a manifold by analyzing the functions on it.

neighborhood   If x is a point in a topological space X, then a neighborhood of x is an open set containing x.

neighborhood base   If $x \in X$ is a point of the topological space, then a neighborhood base is a family of neighborhoods such that every neighborhood of x contains some element of the family.

net   A generalization of a sequence, where a net is indexed over a directed set.

nondegenerate critical point   A critical point for the Morse function h at which the Hessian of h has nonzero determinant.

normal space   A T_4 space in which points are closed.

one-point compactification   A compactification that is created by adding a single point to a space.

open covering   A covering for a set that consists of open sets.

open set   One of a collection of distinguished sets in a topological space.

paracompact space   A space for which every open cover admits a locally finite refinement.

partition of unity   A collection of functions on a manifold that sums to 1. Usually a partition of unity is subordinate to an open cover.

path-connected space   A space in which any two points can be joined by a path.

pointwise convergence topology   If $\mathcal{F}(S)$ is a space of real-valued functions with domain S and $s \in S$, if U is an open set in $\mathbb{R}$, then a subbasis for this topology is sets of the form $\mathcal{G}_{s,U} = \{f \in \mathcal{F} : f(s) \in U\}$.

product topology   The topology on a product space that is generated by the topologies of the component spaces.

proper mapping   A mapping with the property that the inverse image of a compact set is compact.

quotient space   The collection of equivalence classes induced by an equivalence relation.

quotient topology   If $f : X \to Y$, then the quotient topology on Y is a topology induced by the mapping f and the topology on X.

regular space   A T_3 space in which points are closed.

relative topology   If Y is a subset of a topological space X, then the relative topology on Y is a topology induced by the topology on X by way of intersection.

retract   If X is a topological space and $A \subseteq X$ then A is a retract of X if there is a mapping of X to A that fixes the points of A.

retraction   The mapping that makes A a retract of X.

second countable   A topological space that has a countable basis.

separable space   A topological space that has a countable, dense subset.

separation axioms   Axioms for the structure of a topological space. The most common separation axioms are T_0, T_1, T_2, T_3, and T_4.

simply connected space   A space with trivial first homotopy.

Sorgenfrey line   A special topology on the real line that gives an example of a normal topological space whose product with itself is not normal.

sphere   The set of points in space that are distance 1 from the origin.

stereographic projection   A device for compactifying the complex plane.

Stone-Čech compactification A compactification based on the Tychanoff embedding theorem.

subbasis A family of open sets that generates the entire topology by way of finite intersection and arbitrary union.

subcovering A subcollection of a given covering that is itself a covering.

subordinate A partition of unity is usually subordinate to an open cover, which means that each function in the partition is supported in one element of the cover.

support of a function The set on which the function is nonzero.

T_0 **space** The space X is T_0 if, whenever $P, Q \in X$ are distinct points, then either there is a neighborhood U of P such that $Q \notin U$ or else there is a neighborhood V of Q such that $P \notin V$.

T_1 **space** The space X is T_1 if, whenever $P, Q \in X$ are distinct points, then there are neighborhoods U of P and V of Q such that $Q \notin U$ and $P \notin V$. It is easy to check that, in a T_1 space X, any singleton set $\{x\}$ will be closed.

T_2 **space** The space X is T_2 (also called *Hausdorff*) if, whenever $P, Q \in X$ are distinct points, then there are neighborhoods U of P and V of Q such that $U \cap V = \emptyset$.

T_3 **space** The space X is T_3 (also called *regular* if points are closed) if, whenever $P \in X$ and $F \subseteq X$ is a closed subset not containing P, then there are a neighborhood U of P and V of F so that $U \cap V = \emptyset$.

T_4 **space** The space X is T_4 (also called *normal* if points are closed) if, whenever E and F are disjoint closed sets in X, there are neighborhoods U of E and V of F such that $U \cap V = \emptyset$.

topological invariant An entity, usually algebraic in construction, that is invariant under homeomorphism.

topologist's sine curve An important example in the theory of connectedness.

topology The collection of open sets in a space.

torus The product of the circle with itself.

totally disconnected space A space in which the connected components are points.

trivial uniformity The smallest uniformity on a space.

Tychanoff embedding theorem A theorem about embedding a T_1 space into a product of spaces by way of point evaluation.

Tychanoff space A completely regular T_1 space.

Tychanoff's theorem The result that the product of compact spaces is compact.

uniform approximation Let g be a real-valued function on a space X. We say that g is uniformly approximated by functions f_j if, for each $\varepsilon > 0$, there is an N so large that if $j > N$ then $|g(x) - f_j(x)| < \varepsilon$ for all $x \in X$.

uniformity A topology that gives a space a uniform structure.

uniformity of uniform convergence A uniformity that facilitates the definition of uniform convergence of functions.

uniformly convergent sequence of functions A convergent sequence of functions that is a Cauchy net relative to a uniformity of uniform convergence. In the context of a metric space (Y, e), a sequence $f_j : X \to Y$ is uniformly convergent to g if, for each $\varepsilon > 0$, there is an $N > 0$ such that if $j > N$ then $e(f_j(x), g(x)) < \varepsilon$ for all $x \in X$.

uniform space A topological space equipped with a diagonal uniformity.

Urysohn's lemma A result about the existence of a continuous separating function in a normal space.

Weierstrass approximation theorem The result that says that a continuous function on an interval $[a, b] \subseteq \mathbb{R}$ can be uniformly approximated by polynomials.

Weierstrass nowhere differentiable function A continuous function on the unit interval $[0, 1]$ that is not differentiable at any point of the interval.

Bibliography

[ADF] C. Adams and R. Franzosa, *Introduction to Topology: Pure and Applied*, Prentice-Hall, Upper Saddle River, NJ, 2008.

[ARM] M. A. Armstrong, *Basic Topology*, Springer, New York, 1983.

[DAV] A. Davis, *Gödel's Theorem*, University of Oklahoma preprints, 1964.

[GRH] M. J. Greenberg and J. Harper, *Algebraic Topology: A First Course*, Benjamin/Cummings, Reading, MA, 1981.

[HU] S. Hu, *Elements of General Topology*, Holden-Day, Inc., San Francisco, CA, 1964.

[HUW] W. Hurewicz and H. Wallman, *Dimension Theory*, Princeton University Press, Princeton, NJ, 1941.

[JEC] T. Jech, *The Axiom of Choice*, North-Holland, New York, 1973.

[KAR] R. Karp, The probabilistic analysis of some combinatorial search problems, *Algorithms and Complexity (Proc. Sympos., Carnegie-Mellon Univ., Pittsburgh, Pa. 1976)*, 1-19; Academic Press, New York, 1976.

[KEL] J. L. Kelley, *General Topology*, Van Nostrand, Princeton, NJ, 1955.

[KKM] E. Khalimsky, R. Kopperman, and P. R. Meyer, Computer graphics and connected topologies on finite ordered sets, *Topology and its Applications* 36(1990), 1–17.

[KIS] C. Kiselman, Digital Jordan curve theorems, in Borgefors, G., Nyström, I., and Sanniti di Baja, G., eds., *Discrete Geometry for Computer Imagery*, DGCI 2000 Proceedings, Lecture Notes in

Computer Science, Volume 1953, New York: Springer-Verlag, 2000, 46–56.

[KOKM] T. Y. Kong, R. Kopperman, and P. R. Meyer, A topological approach to digital topology, *American Mathematical Monthly* 98(1991), 901–917.

[KRA1] S. G. Krantz, *Real Analysis and Foundations*, 2nd ed., CRC Press, Boca Raton, FL, 2005.

[KRA2] S. G. Krantz, *The Elements of Advanced Mathematics*, 2nd ed., CRC Press, Boca Raton, FL, 2002.

[KRP1] S. G. Krantz and H. R. Parks, *A Primer of Real Analytic Functions*, 2nd ed., Birkhäuser Publishing, Boston, MA, 2002.

[KRP2] S. G. Krantz and H. R. Parks, *The Implicit Function Theorem*, Birkhäuser Publishing, Boston, MA, 2002.

[MIL] J. Milnor, *Morse Theory*, Princeton University Press, Princeton, NJ, 1963.

[MUN] J. Munkres, *Topology*, Prentice-Hall, Upper Saddle River, NJ, 2000.

[NAN] E. Nagel and J. R. Newman, *Gödel's Proof*, New York University Press, New York, 1958.

[NMP] National Mapping Program Technical Instructions, preprint.

[ROS] A. Rosenfeld, Digital topology, *American Mathematical Monthly* 86(1979), 621–630.

[RR1] H. Rubin and J. Rubin, *Equivalents of the Axiom of Choice*, 1st ed., North-Holland, Amsterdam, 1963.

[RR2] H. Rubin and J. Rubin, *Equivalents of the Axiom of Choice*, 2nd ed., North-Holland, Amsterdam, 1985.

[RUD] M. E. Rudin, A new proof that metric spaces are paracompact, *Proc. AMS* 20(1969), 603.

[RUDW] W. Rudin, *Principles of Mathematical Analysis*, 3rd ed., McGraw-Hill, New York, 1976.

[SHO] J. Shoenfield, *Mathematical Logic*, Addison-Wesley, Reading, 1967.

[SKR] G. F. Simmons and S. G. Krantz, *Differential Equations: Theory, Technique, and Practice*, McGraw-Hill, New York, 2006.

[SMU] R. Smullyan, *Gödel's Incompleteness Theorems*, Oxford University Press, New York, 1992.

[SPA] E. Spanier, *Algebraic Topology*, Springer, New York, 1981.

[STO[R. R. Stoll, *Sets, Logic, and Axiomatic Theories*, W. H. Freeman and Company, San Francisco, 1961.

[SUP] P. Suppes, *Axiomatic Set Theory*, Dover Publications, New York, 1972.

[WIL] S. Willard, *General Topology*, Addison-Wesley, Reading, MA, 1970.

INDEX

About the Author

Steven G. Krantz was born in San Francisco, California in 1951. He received the B.A. degree from the University of California at Santa Cruz in 1971 and the Ph.D. from Princeton University in 1974.

Krantz has taught at UCLA, Penn State, Princeton University, and Washington University in St. Louis. He served as Chair of the latter department for five years.

Krantz has published more than 50 books and more than 150 scholarly papers. He is the recipient of the Chauvenet Prize and the Beckenbach Book Award of the MAA. He has received the UCLA Alumni Foundation Distinguished Teaching Award and the Kemper Award. He has directed 17 Ph.D. theses and 9 Masters theses.